T0327984

The Ultimate Guide to
HOUSEPLANT
PROPAGATION

Quarto.com

© 2025 Quarto Publishing Group USA Inc.
Text © 2025 Lindsay Sisti

First Published in 2025 by Cool Springs Press, an imprint of The Quarto Group,
100 Cummings Center, Suite 265-D, Beverly, MA 01915, USA.
T (978) 282-9590 F (978) 283-2742

Cool Springs Press titles are also available at discount for retail, wholesale, promotional, and bulk purchase. For details, contact the Special Sales Manager by email at specialsales@quarto.com or by mail at The Quarto Group, Attn: Special Sales Manager, 100 Cummings Center, Suite 265-D, Beverly, MA 01915, USA.

29 28 27 26 25   1 2 3 4 5

ISBN: 978-0-7603-9040-5

Digital edition published in 2025
eISBN: 978-0-7603-9041-2

Library of Congress Cataloging-in-Publication Data is available.

Design: Emily Austin, The Sly Studio
Cover Image: Kat Schleicher Photography
Page Layout: Emily Austin, The Sly Studio
Photography: Kat Schleicher Photography and Lindsay Sisti
Illustration: Esme Lintin
Horticultural Proofreader: Leslie F. Halleck, Halleck Horticultural, LLC

Printed in China

The Ultimate Guide to
HOUSEPLANT PROPAGATION

STEP-BY-STEP TECHNIQUES FOR MAKING MORE HOUSEPLANTS . . . FOR FREE!

LINDSAY SISTI
from @alltheplantbabies

COOL
SPRINGS
PRESS

In loving memory of all the houseplants I unintentionally killed when first starting out. I hope this book about creating more plants makes up for it.

Pour a glass of propagation water out for my planties . . .

Contents

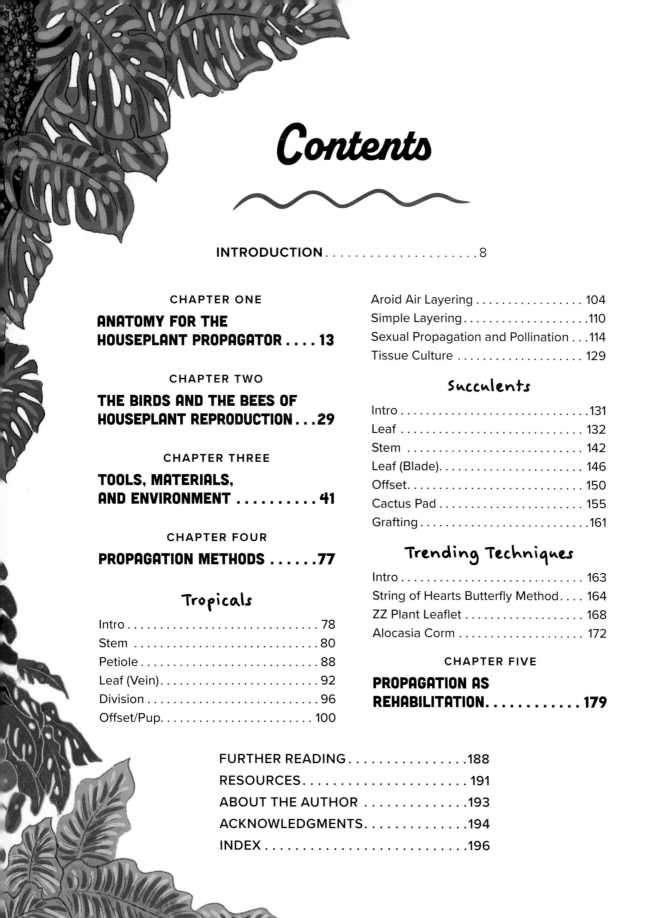

INTRODUCTION

"The best things in life are free."

You know the saying. But if you thought it was about love, happiness, and friendship, you were wrong. It is clearly about plants (the best things in life) and propagation (how to make more, for free). Lucky for you, this book illustrates how to create as many free houseplants as your heart desires. The only thing more rewarding than growing a single houseplant to its fullest potential is propagating it to make *many* houseplants that you grow to their fullest potential. And the best part is, depending on what you do with your new plant babies, the result of all your work may in fact be an increase in love, happiness, and friendship in your life.

Personally, I'm dangerous with a pair of scissors. If you see me walk into a room holding a pair, hide your pothos, hide your monstera, and definitely hide all your glass jars (which, after reading this book, you will lovingly refer to as your propagation vessels). Every time I look at a plant, I immediately see the potential for how many new plants it could become or produce.

I started propagating houseplants as a fun way to share my plants with others and grow my collection, and it slowly evolved into a full-time job. Today I run an online plant shop, and my inventory consists entirely of plants I propagate myself. I continuously work on the creation of new plant hybrids (something I'll teach you how to do in this book), and I have a patent on a new hybrid *Alocasia* species. When I'm not selling or creating new plants, I am filming educational plant content for my @alltheplantbabies social media accounts. I am honored to be able to share my hobby-turned-career and secrets of the trade with you through this book.

The author, Lindsay, of All the Plant Babies, pollinating her *Anthurium carlablackiae* to make anthurium babies

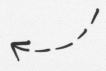

Warning: after reading this book, you may turn into Edward Scissorhands.

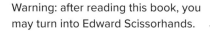

Plant propagations are an environmentally friendly way to communicate your feelings.

WHY SHOULD YOU CHOP AND PROP YOUR HOUSEPLANTS?

Before we get into the step-by-step instructions, let's first review a few reasons why you would want to propagate your houseplants in the first place. There's a lot in this for you!

1. GROW YOUR COLLECTION

Do you want to merely own a single succulent that you named Bob, who sits alone in a window?

No. You want to have an Instagram-worthy, indoor jungle!

Okay, maybe some of you *do* just want to learn how to simply care for a Bob. Whatever your intention, throughout this book we are going to learn how to "chop and prop" and pollinate until you have the indoor plant situation you dream of–anything on the spectrum of ensuring Bob looks fresh over the long term to creating an indoor rainforest!

2. MAKE BACKUP PLANT COPIES

It never hurts to have an heir and a spare. Sometimes when I buy a plant I truly love, I will propagate it right away to make an extra pot in case the mother ship goes down. Making a clone of your plant while it is healthy is a great way to protect against complete decimation of a certain species from your indoor jungle as a result of user error, a pest outbreak, root rot, desiccation, or a fungal/bacterial issue. I made backup plants frequently in my first year as a plant parent in case one of my plants went down faster than the *Titanic*. I knew not to trust my own plant care and rehabilitation skills yet. If you have (or had) a fiddle leaf fig or alocasia, you know how fast it can happen! Use the information in this book to create as many spare copies of your plants as you need to feel secure. Works better than iCloud.

3. GIVE EPIC GIFTS

Plant propagations make fabulous, low cost, meaningful gifts. Never again will you be without the perfect present for a special occasion. Nothing says, "Welcome to your new home!" better than a beautiful plant homeowners can use to start their own jungle. How about sending a "Hoya doin'?" card with some hoya cuttings to an old friend you'd like to catch up with? Propagate your cactus with a "You're so sharp!" card for a new grad. In a bad breakup? Nothing says "you succ" like a succulent! (Better yet, in that scenario, simply gift a piece of a rotting succulent, but for the record, you didn't get that idea from me.) There is truly no limit to the messages you can communicate with your propagations.

A Propagation Glow-up

After your plant cuttings take root, you can replant them back into the mother plant's pot to make it lush. This is common practice with trailing plants.

[left] Before propagation: thin "hair," a.k.a. variegated string of hearts *Ceropegia woodii* 'Variegata'

[right] After propagation, she's runway ready.

4. TRADE THEM FOR SOMETHING NEW

While gift giving is encouraged, plants cost money and you have to pay the bills! How does one afford to sustain the hobby over time? One way is by trading plant specimens. Often times, a plant collector will purchase a plant, grow it out a bit, take cuttings, and then trade the cuttings at a local plant swap or meetup for another plant or cuttings of similar value.

Search on Facebook to see if there is a houseplant or rare plant swap group near you. If you are near a major city, chances are there is at least one. If not, consider starting one! In addition to increasing your collection, joining a plant swap group is a great way to meet other amazing plant people (who else will be able to empathize with you about mealybugs?), as well as learn more about the hobby. If there is no local meet up group in your area, there are national B/S/T (Buy, Sell, Trade) groups for various types of plants on Facebook and often people swap by mailing each other plants across the country. I once traded a few rare plants for a custom ring from a jeweler I followed on Instagram. Plants are not just the new pets—they're the new currency! Keep in mind, though, that if you're selling plants commercially, most countries and states have regulations involved and you must have the proper licenses.

5. SELL THEM ~~TO PAY THE BILLS~~ FOR MONEY TO BUY MORE PLANTS!

Another common practice in the plant collecting community is to sell plant propagations either rooted or unrooted to recoup at least part, if not all, of the cost of the mother plant. Some avid collectors like me eventually end up opening their own online or brick-and-mortar plant shops rooted from their propagations (see what I did there)? Because healthy plants continue to regenerate, they are the shop inventory that keeps on growing, literally.

If you are thinking about selling plant propagations for profit, however, set your expectations to match the current market's prices. The post-pandemic plant market is saturated, and prices of plants have plummeted overall. During the plant boom of the pandemic in 2020, many sellers were making tremendous profits from plant cuttings and propagations. Online and brick-and-mortar plant shops popped up left and right. To meet this demand and also capitalize on the trend, tissue culture laboratories (more on this in chapter four) that mass produce small starter plants in laboratories ramped up production as well.

Then, our world changed once again. The pandemic came to an end and people were no longer quarantined at home. They starting putting houseplants on the backburner as they ventured back out into the world (the audacity!). Demand dropped when supply was high, and so prices plummeted. Amazing for plant collectors, not so much for brick-and-mortar plant shops with steep overhead costs.

One thing that remains today, however, is the large variety of social media and e-commerce platforms available for collectors and hobbyists to buy, sell, and trade plant propagations. Etsy, Shopify, Facebook, and Instagram are e-commerce platforms and social media networks from which to sell plants and/or host online shops, to name a few. And while starting an official plant shop may not be as profitable as it was during the pandemic, with the right mindset and pricing strategies, one can be successful. My Etsy shop, which focuses primarily on anthuriums that I hybridize, is still profitable after three years.

Propagation helps to reduce the expense of having to purchase your entire store's plant inventory from a wholesale nursery or import the plants from overseas. Growing inventory through propagation is more work and slower, but it helps to increase profitability.

6. REHABILITATE A STRUGGLING PLANT

To do any of these wonderful things I just spoke about with your cuttings, however, you must maintain robust and healthy "mother plants" that stand the test of time. You may not know this, but after buying a plant, the credit card receipt is a symbol of your union with said plant. It signifies that you've agreed to take care of it through sickness and in health—and rehab, if it starts to rot.

Root problems, fungal and bacterial issues, and pests are no strangers to our chlorophyll-filled friends. It's how well equipped you are to help your plants overcome them that will determine whether your plants survive long term and continue to produce new growth or end up in the compost. While the treatment of these individual issues is beyond the scope of a book focused on propagation, in chapter five we will discuss how to perform plant surgery to rescue a struggling plant and grow it anew. Using plant propagation as a way to rescue a plant is perhaps the number one most important reason to master this art.

Whatever your reasons for embracing the hobby of houseplant propagation, let's begin by meeting the anatomical parts of plants involved in propagation.

My online plant shop's home-grown inventory

Chapter 1

ANATOMY FOR THE HOUSEPLANT PROPAGATOR

What did the pothos cutting dress up as for Halloween?
A propa-ghoul!*

A propagule is any structure that can give rise to a new individual organism. In plants, this includes stems, corms, tubers, seeds, and spores.

Before surgeons are allowed to operate on a human body, they must first become a master of anatomy. Likewise, to propagate houseplants correctly, you need to be familiar with basic plant anatomy. If you chop your plant's stem in the wrong place, you're not likely to kill your entire plant. However, you may be left with an ugly nub or a cutting that doesn't grow roots. To prevent this terrible fate, let's start by reviewing the plant parts you need to know in order to make plant babies.

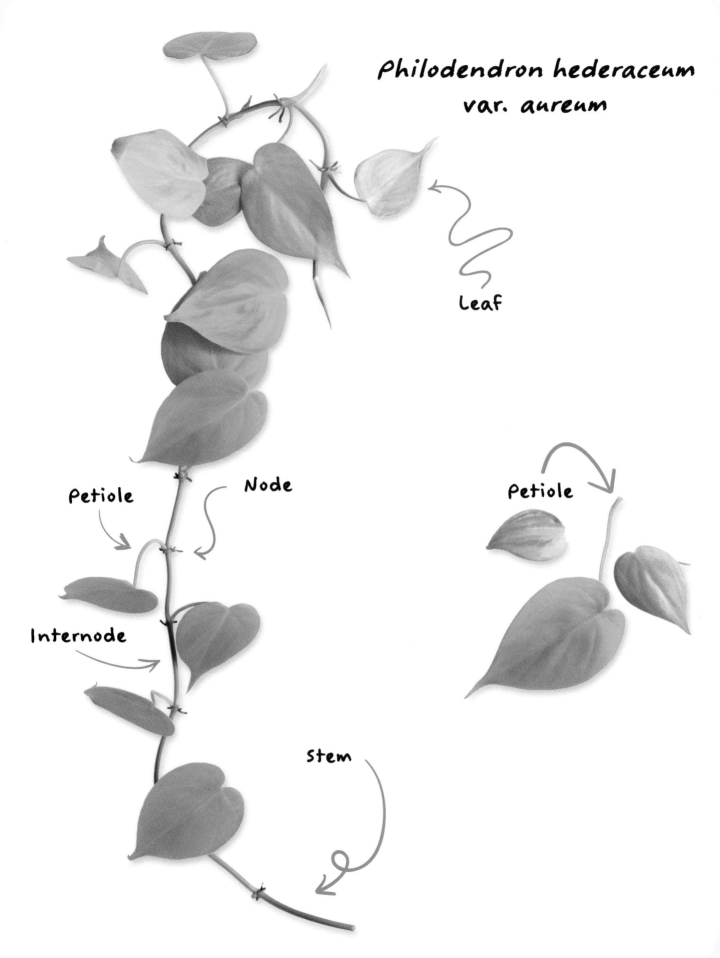

Philodendron hederaceum
var. *aureum*

Leaf

Petiole

Node

Internode

Stem

Petiole

Beginner's Houseplant Propagation Anatomy featuring *Philodendron hederaceum var. aureum*

Leaves

Leaves are the primary sites of **photosynthesis**. Photosynthesis is the process in which plants use sunlight, water, and carbon dioxide to create oxygen and energy in the form of sugars.

Stems vs. Petioles

Many people think that every leaf-bearing structure on a plant is called a stem. However, the stem is only the main, thicker axis of the plant. It acts as the highway through which water, food, and minerals are transported to other parts of the plant. Additionally, green stems, like the other green parts of the plant, create plant food through photosynthesis. Petioles, however, are thinner and are the stalk *off* of the stem. They connect the stem to the base of a leaf.

Thin stalk with leaf attached = Petiole

Thick central axis = Stem

When it comes to propagation, knowing the difference between the stem and petiole is crucial to your success. Petiole propagation is when you cut the leaf off at the base of its petiole and then encourage the petiole to grow its own roots and shoots (see page 88). Some plants will fully propagate this way, some will simply grow roots from the petiole but won't grow shoots (looking at you, rubber plant), and some won't do a damn thing unless you propagate the stem or leaf.

So, in this book, when the instructions are to "cut the stem," cut the thick, central part of the plant with the main character energy, okay? And when the instructions are to "cut the petiole," don't get pet-i-ol-bout it. (Petty-about . . .okay, that was a stretch.)

A petiole cutting that produces roots but never shoots, like this ruby rubber plant (*Ficus elastica* 'Ruby') is what I call a tease. Horticulturally speaking, it's called a blind cutting. To properly propagate a rubber plant, take a stem cutting (cut the main branch of the plant that holds the petioles and leaves, as discussed on page 78), or air layer it (see page 104).

Nodes

If the main stem is a highway for water and nutrients, nodes are the rest stops where you can get Starbucks and a Slim Jim. Nodes are sites that occur at intervals along your houseplant's stem from which new growth can arise. Typically, they appear as notches or bumps along the plant's stem, and on some plants, you'll see aerial roots emerging from the area. When taking a cutting, always cut below a node. And if you didn't node that, now you node.

Internodes

The spaces along the stem, between the nodes.

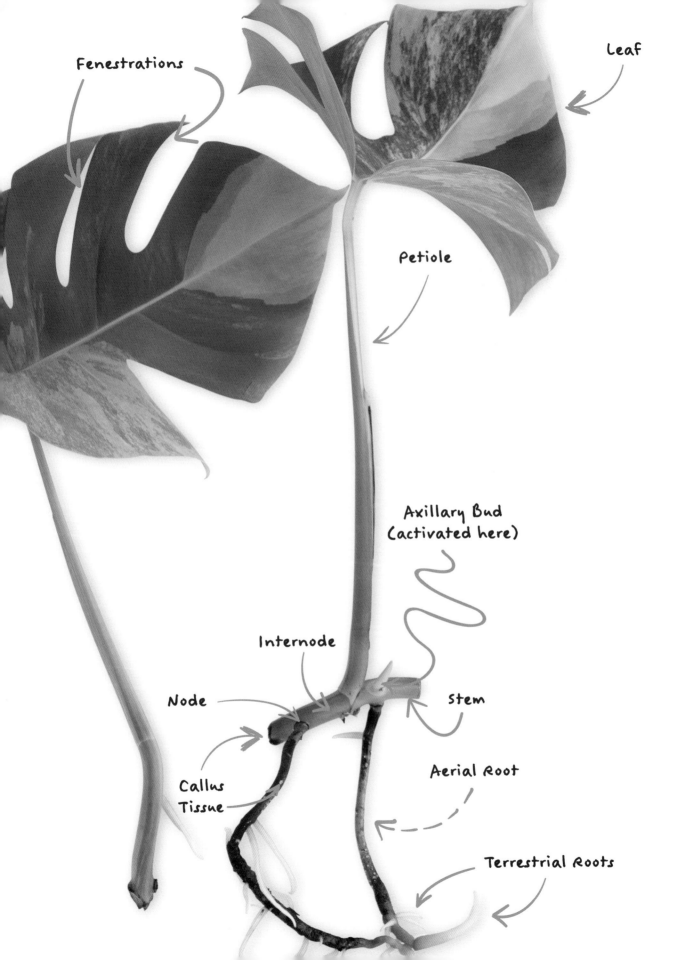

Fenestrations

Leaf

Petiole

Axillary Bud
(activated here)

Internode

Node

Stem

Callus
Tissue

Aerial Root

Terrestrial Roots

Aroid Propagation Anatomy
Featuring *Monstera deliciosa* 'Aurea'

Fenestrations

Referred to as "fennies" in the hipster horticultural crowds, fenestrations are naturally occurring holes in leaves of certain plant species such as *Monstera deliciosa* and *Monstera adansonii*. It is believed that fenestrations are adaptations to help the mature plants to withstand gusts of wind without damage as they climb up the sides of trees to great heights in the forest canopy. They're also fun to peek through.

Axillary Bud

Often appearing as a little bump, an axillary bud is an area of growth on a stem of a plant, usually where a petiole meets a stem. Axillary buds will remain dormant until the plant activates them, and when it does, some will grow into new shoots. How can you purposefully activate the axillary buds on your plant to encourage the plant to produce more lateral shoots/branches?

Chop the top off!

This is why beheading your fiddle leaf fig will often cause the base of the plant to grow back with more branches than before. Don't get too excited though; only a small number of axillary buds will typically activate at one time. The word *axillary* stems from the word *axil*, the part of the leaf the bud emerges from. It is not to be confused with the word *auxiliary*, which means supplementary. I may or may not have confused the two words for half a decade.

Callus Tissue

Don't freak out if you propagate a cutting in a jar of water and firm, white bumps form around the stem and/or roots. Chances are, you do not have a sudden mealybug invasion, but callus tissue. Generating callus cells is the plant's natural way of healing in response to stress (such as being cut). It's the first step in regeneration into a new plant.

This tissue is an **undifferentiated** mass of cells. This means the cells are not assigned a specific task. They can become roots, shoots, or nothing at all when they grow up, depending on environmental factors, what rooting hormones are used (see page 73), and if you're a strict parent or a proponent of gentle plant parenting techniques. But it's not necessary to scrape the callus off. Just leave it and let the plant do its badass regenerative routine.

The term callus may sound familiar to you, because often, when propagating plant cuttings, people will advise that you "let the cutting callus" prior to planting it in substrate. They are implying that you let the cut area dry up a bit and seal the wound you made with callus tissue. Some of the tissue will go on to become new roots and shoots. Callus tissue looks different during different types of propagation, however, and is usually barely noticeable. It's only during water propagation that I've witnessed it turn thick and white.

Aerial Roots

When deciding where along a plant's stem to take your cutting, notice first if there are aerial roots (also called **adventitious roots**) present. New roots will form from the existing aerial roots first during propagation. Aroids, such as monsteras and philodendrons, and many vining plants, such as hoyas, grow up the sides of trees in the wild. The more humid your home environment, the more prominent the aerial roots will be on these plants. In areas of high humidity, tropical plants stretch out their "limbs" (a.k.a. aerial roots) in hopes of finding a tree to attach to and climb up, as it would in the jungle. Just as would happen if an aerial root touched the jungle floor in the wild, during plant propagation in humidity enclosures, aerial roots often transform into underground/terrestrial roots!

Terrestrial Roots

Roots that grow underground. Is there an echo in here?

[right] White callus tissue on the bottom of a rubber tree cutting. Remember though—callus is without malice! Just leave it be.

[page left] *Monstera deliciosa* 'Aurea'

SPECIALIZED STEMS

*"When a plant walks in with an itty bitty stem and a round thing in your face—
You get sprung."*
—Baby Got Bulbs

Alocasia 'Green Unicorn' showing off its specialized stems

Plants are resilient organisms. They've evolved to survive periods of harsh environmental conditions, such excessive heat, too little light, and periods of drought. This works out in our favor because when we bring them inside as houseplants, many of them are built to survive brief periods of indoor environmental stress or human neglect. This includes putting them next to cold drafts, leaving them in excessively hot places, forgetting to water them, stress from shipping, surviving your dog's farts, etc.

Some plants do this by either entering a state of genetically controlled true dormancy, where the roots, leaves, and stems whither down to the plant's underground storage structure and all cell division stops. Other plants, including the tropical plants we grow as houseplants, do this by entering a resting period where active growth stops and the plant may or may not die back but cellular growth and division does not cease. These resting periods are typically triggered by exposure to certain temperatures, shortened day length, a lack of water, or other stressors. Both adaptations serve plants well during dry and rainy or cold and hot seasons in the wild—and during busy life periods for innocently neglectful plant parents. Unfortunately, there is nothing that can help them survive wild human children however, so use protection. As in, protect your plants (sheesh).

Bulbs, such as onions, are the most commonly recognized underground plant storage structures. All true bulbs store food for the plant and are enclosed in scales. However, in this book we are going to skip a detailed discussion about true bulbs because you're not here to propagate onions. Rather, we are going to take a close look at the houseplant underground storage structures you are most likely to encounter as a houseplant propagator: rhizomes and corms and tubers. Oh my!

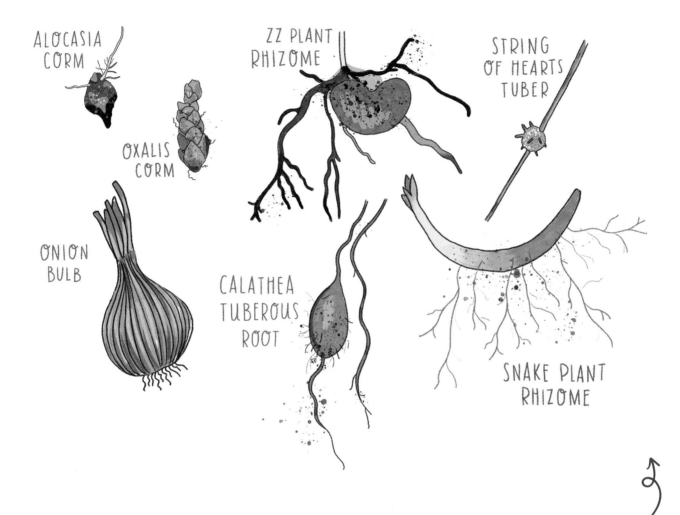

ALOCASIA
CORM

OXALIS
CORM

ZZ PLANT
RHIZOME

STRING
OF HEARTS
TUBER

ONION
BULB

CALATHEA
TUBEROUS
ROOT

SNAKE PLANT
RHIZOME

Underground plant structures are the plant version of underground survival bunkers. It's important to note that only some houseplants contain these storage structures. If a plant does not have one and it starts to struggle, don't fret! You can still save it through stem or leaf propagation, but more on that later.

What do plants keep in their survival storage structure that is keeping them alive during these rest periods? Hint: it's not canned beans and N95 masks. In each corm, tuber, and rhizome, a plant stores carbohydrates and nutrients it can convert into energy when needed.

Common houseplants that can enter a rest period include alocasia, purple shamrock plant (*Oxalis triangularis*), and snake plants (*Dracaena* spp.). As long as a plant's storage structure is still intact without rot, there is a good chance you can bring it back into an active growing state with fresh substrate, ideal growing conditions, and water to revitalize it.

Common houseplant storage structures (*plus a stray onion bulb for comparison*)

CORM

A corm is an underground modified stem that stores food for certain plants such as *Oxalis triangularis*. Oxalis corms start off tiny, but over time grow long and skinny with scales on the outside. They eventually look like a brown version of those Chinese finger traps you may have played with as a kid.

Corms are different than bulbs in that they're solid inside and sometimes are referred to as solid bulbs by the cool kids. Unlike bulbs, when you cut a corm in half, you won't see concentric rings, like you would on a bulb (picture the rings on an onion sliced in half). Roots grow from the base of the corm, and shoots will grow from the top, so it's important to place them right side up when planting them.

Oxalis Corm

Alocasia Corm

Alocasias produce corm-like storage structures that protrude from underground stems. They are most commonly referred to as *corms* in the United States and *bulbs* outside of the United States. However, their true botanical descriptor remains a heated debate among people who enjoy debating such things.

To make matters a bit more confusing, alocasias produce these storage structures in different shapes and sizes and at different locations along the plant, depending on the species. Some botanists, scientists, and hobbyists refer to these structures as **bulbils** (structures that look like bulbs but aren't), and both the plant taxonomist and biologist Peter Boyce and the Royal Botanic Gardens, Kew refer to them as **tubercules**, which are small knobby prominences, or nodules, on a plant. Other botanists describe them as **cormels**.

One thing is widely agreed upon, however; if you use the phrase *Alocasia corm*, everyone will know you are referring to a small bulb-like structure under the soil connected to the plant, because that is the term most widely used throughout the hobby. Therefore, throughout the rest of the book, I will also use the term *Alocasia corm*.

TUBER

Another modified stem with energy-storing super powers is the tuber. Roots and shoots grow from buds throughout its surface. By far, the most famous and tastiest tuber to ever exist is the potato, but again, we are not here to discuss my love of french fries but instead houseplant propagation.

Let's take a closer look at the dainty string of hearts (*Ceropegia woodii*), for example. Not only does this beautiful plant grow from tubers under the soil, but it also forms round "aerial tubers" sporatically along more mature vines. In the wild in its native southern Africa, as the plant vines down trees and rocky outcrops in the landscape, these aerial tubers form new roots down into the ground wherever they make contact with the earth. New string of hearts vines will form from there. You can mimic this happening in your home through a propagation technique called **simple layering** (see page 110). Sadly, however, unlike potatoes, you can not fry these tubers up to make crispy fries.

String of Hearts Tubers

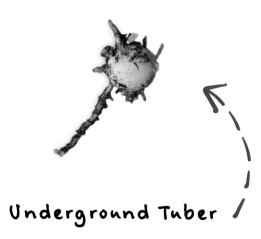

Aerial Tuber   Underground Tuber

RHIZOME

Rhizomes are specialized storage stems that typically grow horizontally at or just below the soil surface and typically connect clumps of plants that you can divide apart during propagation. They come in many shapes, colors, and sizes.

The rabbit's foot fern is unique in that it has above ground, fuzzy tarantula-leg-looking rhizomes. You can actually propagate rabbit's foot fern by cutting off sections of the fuzzy rhizomes and pinning them down accross the top of damp potting mix. As long as it is given adequate light and the potting mix is kept damp, new roots and shoots should grow from the cute fuzzy rhizomes.

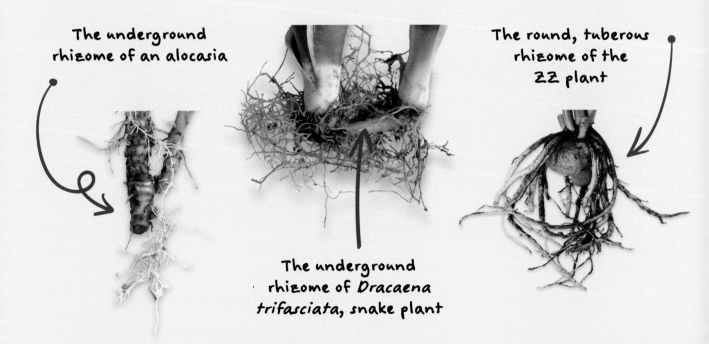

The fuzzy aboveground rhizome of the *Davallia fejeensis*, rabbit's foot fern

The underground rhizome of an alocasia

The round, tuberous rhizome of the ZZ plant

The underground rhizome of *Dracaena trifasciata*, snake plant

A new, baby ZZ plant rhizome enters the world through water propagation.

STOLON

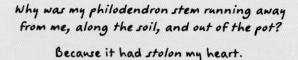

Why was my philodendron stem running away
from me, along the soil, and out of the pot?

Because it had *stolon* my heart.

Stolons are horizontal stems that typically run along the top of the ground, or just beneath it, take root and produce new plants at certain points along their length. Stolons can be either super thick and heavy, as is the case with creeping philodendrons such as *Philodendron gloriosum* and *Philodendron plowmanii*, or they can be thin and lightweight, as they are with strawberries. When growing creeping philodendrons as houseplants, it is ideal to grow them in long, rectangular pots to accommodate their habit of roaming along the earth like a groundcover. Plants with stolons are called stoloniferous. Plant moms who grow plants that have stolons are called babelicious.

Telling the difference between a rhizome and a stolon can be tricky. For me personally, it helps to remember that creeping philodendrons have stolons, while most of my other houseplants with similar looking storage structures have rhizomes.

[top] My mature *Philodendron plowmanii* plant

[bottom] The same *Philodendron plowmanii*'s stolon grows along the top of the potting mix.

A *Philodendron pastazanum*'s aboveground stolon creeping along the soil vs. a snake plant's belowground rhizome. Both stolons and rhizomes are modified stems that enable the plant to spread across the ground.

SEXUAL REPRODUCTIVE PARTS

Why wouldn't the butterfly pollinate the corpse flower (Amorphophallus titanum)?

Because it didn't like the *stigma* attached to it.

It's time to review the plant anatomy that makes young houseplants giggle in sex-ed class! Growing up, and perhaps still, you probably didn't think much about flowers other than the fact that they looked and smelled pretty. If you only knew that flowers contained all the plant's "private parts," you probably would have had a good giggle too.

Flowers are the reproductive parts of plants and contain male and/or female sex organs. The male sex organ of the flower is called the **stamen**. Stamens have little fertile tips that house pollen, called anthers.

The main female sex organ of the flower is called the **pistil**. The typically vase-shaped pistil has three parts:

1. The **stigma**: This is the top of the "vase." This sticky tip attracts pollen grains from the air or a pollinator. It is where pollen lands and germinates. Sticky stigma—easy to remember!

2. The **style:** This is the neck of the vase that leads down to the base.

3. The **ovary**: This is where the action happens. The ovary produces the eggs/ovules. After pollination and fertilization, the ovules will develop into seeds.

Flower structures (a.k.a. morphology) varies among different plant species. Some plants have flowers that are considered **perfect**, or **bisexual** (see flower #1 on the diagram to the right). This means they contain both functioning male and female parts (male stamens and female pistils) and some perfect flowers can reproduce through self-pollination if their own pollen reaches their stamens. Scandalous, I know. Houseplants that have bisexual flowers include peace lilies, hoyas, peperomias, anthuriums, philodendrons, and monsteras.

Other plants produce flowers that are **imperfect**, or **unisexual** (see flowers #2 and #3 on diagram). Each flower is either male or female. Male unisexual flowers that contain stamens (and not pistils) are referred to as *staminate* flowers, and female unisexual flowers contain pistils and are referred to as *pistilate*. Alocasias, philodendrons, and rex begonias all produce separate male and female flowers on a single plant.

I bet now you're looking at all your plants' flowers and wondering their sexuality. It's only natural to want to know!

Three flower configurations

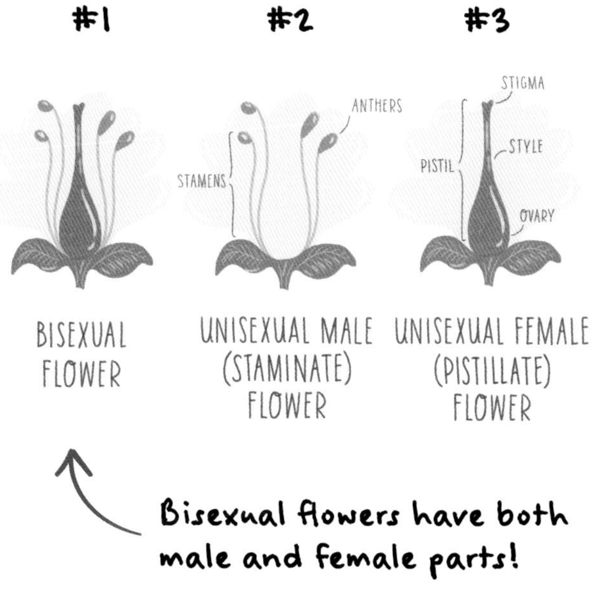

#1   #2   #3

ANTHERS

STAMENS

STIGMA

STYLE

PISTIL

OVARY

BISEXUAL FLOWER

UNISEXUAL MALE (STAMINATE) FLOWER

UNISEXUAL FEMALE (PISTILLATE) FLOWER

Bisexual flowers have both male and female parts!

FLOWER ARRANGEMENTS

Depending on what family a plant is from, a houseplant may produce a single flower at a time, like a bird of paradise, or a cluster of flowers arranged on a stemlike axis, called a **peduncle**. The arrangement of flowers on a peduncle is called an **inflorescence**.

In hoyas, for example, flowers are arranged in inflorescences that look like little umbrellas. Luckily for you, the name for this configuration is easy to remember because it sounds like umbrellas: **umbels**. The short stalks that connect all the flowers together at a point are called **pedicels**.

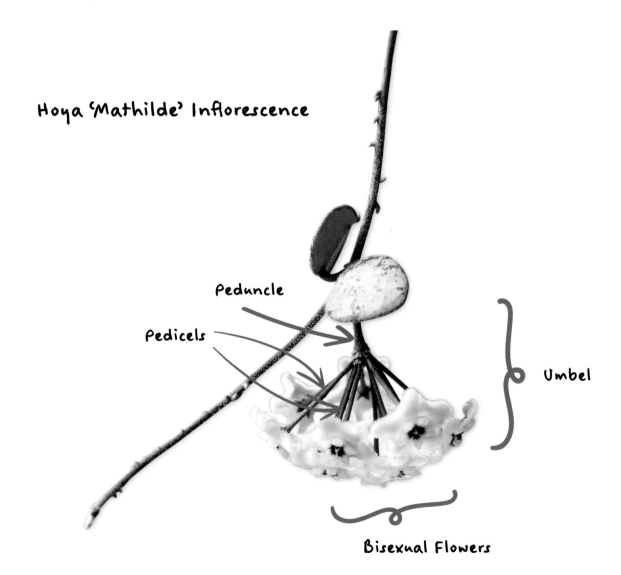

Hoya 'Mathilde' Inflorescence

Peduncle

Pedicels

Umbel

Bisexual Flowers

Alocasia Inflorescence

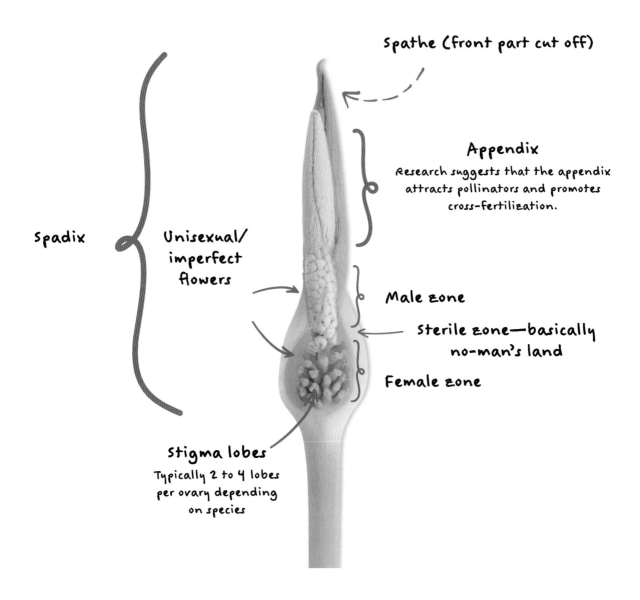

Spathe (front part cut off)

Appendix
Research suggests that the appendix attracts pollinators and promotes cross-fertilization.

Spadix

Unisexual/ imperfect flowers

Male zone

Sterile zone—basically no-man's land

Female zone

Stigma lobes
Typically 2 to 4 lobes per ovary depending on species

Another type of inflorescence, found on plants in the family Araceae (i.e. monsteras, alocasias, anthuriums), is called a **spadix**. A spadix is a spikelike configuration in which the flowers are aligned on what botany books describe as a "fleshy stem." Creepy, but accurate. The spadix is surrounded by what looks like a petal, but it is actually a modified leaf called a **spathe**. Above is an alocasia inflorescence as an example.

I hope this look at basic houseplant anatomy has enabled you to see the propagation potential of your houseplant collection. Now that you know about the plant parts involved in propagation and reproduction, it's time to learn how they work together to create baby plants. You've seen the boats, now let's learn about the motion in the ocean!

Chapter 2

THE BIRDS AND THE BEES OF HOUSEPLANT REPRODUCTION

Simply put, propagation is the process of creating new plant babies. There are two types of propagation: **asexual** and **sexual**. Since this book will cover techniques for both types of propagation, let's take a moment to understand what these terms imply.

A Special Note

Throughout this book I use the term **mother plant** to describe the plant you are propagating because that is the term used in the plant hobby community to indicate a parent plant. This does not imply that the plant or flowers are gender female or that I have daddy issues.

ASEXUAL PROPAGATION

Asexual plant propagation is when you make **clones** of the mother plant using stems, leaves, roots, crowns, or underground storage structures (bulbs, corms, etc.). The new baby plants you make as a result of asexual propagation are genetically identical to the mother plant. Taking a plant cutting and rooting it is an example of asexual propagation. Dividing a plant to make new ones is asexual propagation. Grafting, layering, and tissue culture (see pages 110, 129, and 161), also known as micropropagation, are other methods of asexual propagation. No pollination or plant sex is involved here!

Next time your friend says, "Thanks for the philodendron cutting!"

You can reply, "You're welcome, I enjoyed the asexual propagation."

This type of response will make your popularity skyrocket.

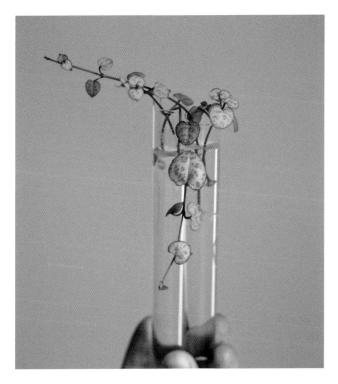

Taking a cutting and placing it in a jar of water is an example of asexual plant propagation.

A *Pilea peperomioides* plant and its pups. You can separate the pups from the mother plant through the simple technique of division, which is a form of asexual plant propagation. Each pup is a genetic clone of the mother plant.

Variegation Variation

Although your plant cutting, pup, or tissue cultured plant is considered a clone of the parent plant, that does not imply that the variegation will be the same as the mother. Not all types of variegation are genetically stable. For example, **chimeric variegation**, the type of variegation you'll find on a variegated *Monstera deliciosa* 'Albo-variegata' (colloquially known as monstera Albo) is the result of genetic mutation and is not stable. When I sell rooted cuttings of my variegated philodendrons or monsteras, many of my customers will ask to "see the mother plant" to scout out what the variegation potential is for the rooted cutting they're about to buy. When I show them, I explain that the variegation of the mother plant is not completely indicative of how variegated the baby plant will be throughout its life. This is because chimeric variegation shows up differently over time in each cutting/clone. You can take a node cutting from a very pink variegated pink princess philodendron (*Philodendron* 'Pink Princess') and it will grow out to be a solid green plant.

Asexual propagation is pretty wild when you think about it. You can just rip off one of your plant's limbs and it grows into an exact clone of the original plant. Imagine if humans could clone themselves in this way! Imagine if you could chop off a finger, place it in water in a nice glass vessel on your windowsill, and it would grow into an exact clone of you. It is possible in plants, but not in people, because some plant cells are **totipotent**, meaning they are able to differentiate, or separate into different parts of a plant and divide until they create an entirely new organism.

The term totipotent is not only fun to say out loud with a British accent, but it is extremely relevant when it comes to **tissue culture** propagation as well, which we'll explore more in chapter four. Tissue culture, also called micropropagation, is a propagation technique in which a large number of plantlets are created from plant tissue in a sterile laboratory environment. Because some plant cells are totipotent, they can grow from clusters of plant cells in a laboratory into full plants that you eventually purchase from the store.

You may (or may not be) wondering: What is preventing humans from reproducing asexually, besides the fact that it's utterly creepy? Why must we be condemned to a life of sexual intercourse and complex emotional relationships for reproduction to occur? Wouldn't the whole finger-in-the-glass reproduction method be much more efficient than having to procreate with someone? I know these questions keep you up at night.

Biologically speaking, the reason why we cannot reproduce asexually is because mature human cells are not totipotent and cannot differentiate and divide to make a new organism. In other words, not every cell in your body contains all the necessary genetic information needed to make a new you. While certain human embryonic cells are totipotent, and can differentiate into various cells, after growing and developing, mature human cells become specialized based on their location and function. Your liver cells will remain liver cells. Your skin cells will remain skin cells. Your finger will remain your finger—even if you put it in a gorgeous propagation jar with rooting hormone powder and a heat mat and Instagram it (so don't try it). Continuing to swipe right or left may still be a better use of your finger.

SEXUAL PROPAGATION

They say sex sells, but will plant sex help sell this book? It's worth a shot . . .

Sexual propagation is essentially just that, plant sex. The result of said sexy time is the production of seeds. Just as with humans, sexual reproduction in plants involves female and male sexual organs, but instead, it typically happens within flowers and is slightly less or more exciting, depending on who you ask. Unlike asexual propagation where you are creating clones, in sexual propagation, just as when two humans make babies, you are mixing up the parent's genes and producing new plants that are genetically different from each of the parent plants. When I grow out a batch of anthurium seeds, each seedling is unique from one another and from each of the parent plants.

Sexual propagation has a few key advantages over asexual propagation. Primarily, in the wild, the creation of new genetic mixes is beneficial if one assumes that high species diversity is preferable. Sexual propagation is also an important way to introduce certain traits, such as disease resistance, into a plant population over time.

Additionally, sexual propagation can be a more efficient means of propagation than asexual propagation. For the flower-happy anthuriums I grow for example, it is more efficient to hand pollinate the flowers, which will produce up to one hundred seeds at a time, than it is to take one or two stem cuttings and try to root them for four months.

Sexual propagation is also what enables the creation of new varieties and hybrids. For tropical plant enthusiasts and growers, the creation of new species makes the hobby more exciting. Hybrids are often more vigorous than their pure-species counterparts. This phenomenon is referred to as *hybrid vigor* and is a result of outcrossing. My entire Etsy shop is based off the sexual propagation of anthuriums and the creation of hybrid plant species. The new alocasia species I created, *Alocasia* 'Green Unicorn' (PP35,554), is more vigorous in its production of corms and hardier than either of its parents, *Alocasia azlanii* or *Alocasia baginda* 'Dragon Scale.'

Finally, sexual propagation is also beneficial in that it helps avoid a phenomenon called **inbreeding depression**. Anyone who has "selfed" (or self-pollinated) a plant (essentially, bred it with itself) has observed that the offspring may be slower to grow and less vigorous overall than if you bred two separate plants together (from the same species or not). Inbreeding depression is when lineages that demonstrate a high level of

Breeding *Alocasia azlanii* and *Alocasia baginda* 'Dragon Scale' resulted in my hybrid *Alocasia* 'Green Unicorn', which is a hardier species of alocasia.

mating between closely related individuals fail to thrive after a few generations and fail more severely as the generations go by. While self-fertilization of one plant may not be terribly harmful to future generations, if it was to be continued over and over, you may notice reduced vigor, reduced number and size of flowers, and fewer seeds. I absolutely observe this with my anthuriums.

With all these benefits, it sounds like sexual propagation from seed is the obvious way to go over asexual propagation. But remember, with sexual propagation, you're dividing up the individual parent's strong combination of genes among the offspring. Just like the kids of two incredibly good-looking celebrities, you never know what you're going to get. Sometimes you get ugly plant kids, and sometimes you get weaker plant kids.

On the contrary, asexual propagation allows you to create exact clones of the strongest mother plant of your choosing. Also, on a practical note, it can be a very difficult to near-impossible task to get some of your houseplants to flower within your home, let alone to pollinate them and get fertilization to occur. Sexual propagation isn't always a realistic avenue for the houseplant collector to pursue when creating more plants. Most of the time, sexual propagation will be left to Mother Nature and asexual propagation is left to the human plant mother (or father). However, there are exceptions to this, such as the easy-to-pollinate anthuriums, which I'll discuss shortly.

WHAT IS POLLEN?

Before you continue on, we are going to play a game. Every time you read the word sperm in the next paragraph, you have to buy a new plant. Ready? Go!

A grain of pollen is essentially a sperm production and delivery pod. More specifically, it's a two-sperm production pod for angiosperms (the flowering plants). This includes most houseplants you may own, such as anthuriums, hoyas, African violets, Christmas cacti, and many others. The next time you sneeze as a result of breathing in too much pollen, remember it's just because you're allergic to plant sperm.

Each grain of pollen has a pretty magical, protective coat. The coat is waterproof, as the biggest risk to the pollen grain on its arduous journey to reach the female flower parts is desiccation, or drying out. It also enables the pollen grain to adhere to the pollinator and then to the sticky stigma (female part of the flower if you remember from the anatomy section). Finally, this technicolor-

pollen dreamcoat enables the pollen grain to recognize the plant upon which the pollen has landed, to prevent self-pollination. Magical indeed!

Pollen grains vary from plant species to plant species in terms of size, color, and surface texture. They even vary in the way that they smell. While the various scents of flowers are cleverly designed to attract specific pollinators, some of the scents are produced by volatile oils, which have additional functions. For instance, some of these oils taste unpleasant and ward off herbivores who may be looking for a tasty snack. Some volatile oils have antibiotic properties and are therefore protective. Finally, the evaporation of these scented volatile oils helps prevent the pollen grains from overheating.

Now, if I haven't already made clear that this tiny sperm manufacturing and transportation pod with a magical coat is cool, there's more. After reaching its destination on another flower, a pollen grain needs to be able to take up water, germinate, and produce a tube that will grow through the female part of the plant to the ovary to find an egg. There is an opening in the pollen wall of anthurium pollen that allows for this tube to grow.

"But what if the cool little sperm pod with its magical coat gets hungry on its trip to the female part of the other flower?" you ask. I understand your concern—I had the same one.

Well, inside the grain of pollen is a supply of starches and sugars that provide energy for its journey (I imagine microscopic peanut butter and jelly sandwiches) and

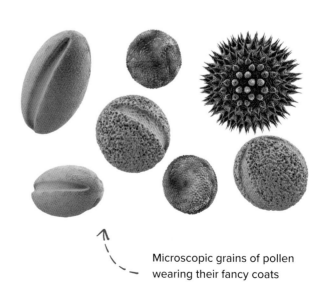

Microscopic grains of pollen wearing their fancy coats

extra energy to produce and feed the sperm (even smaller sandwiches I can only assume; don't dare try to convince me otherwise).

I know, it's just all too cute, and you had NO IDEA this was all going on inside all that yellow dusty stuff. Nature is mind-blowing. Now, go sneeze and blow your nose.

HOW TO AVOID SELFING YOURSELF (AS A PLANT): SEXUAL SYSTEMS AND REPRODUCTIVE STRATEGIES

We know that for most organisms, inbreeding is less than ideal and can promote the passing down of unwanted genetic traits. Most of the time, genetic diversity and adaptation over time to one's environment is ideal.

While it may seem very efficient and convenient to be a bisexual flower that has both male and female sex organs and therefore could theoretically pollinate itself, or a plant that contains a zone of female flowers and a zone of male flowers, and again, could theoretically pollinate itself, self-pollination has both benefits and drawbacks. Life gets a little complicated when you have organs that can produce cells that can fertilize one another. Imagine if you could get yourself accidentally pregnant? There are some people I know that would be in a LOT of trouble.

Some plants have developed various reproductive strategies to avoid self-pollination and promote **outcrossing** (breeding with other plants). One example of a reproductive strategy employed by a plant to avoid self-pollination is that of caudiciforms such as *Stephania suberosa* (you know, the little potato-looking plants). Some *Stephania suberosa* plants are considered male, and others are considered female. The male pollen can be found on the "male" plants and female ovaries and seeds can be found on the "female" plants. They have imperfect, or unisexual, flowers, but they're on separate plants. To reproduce sexually, the entire population of plants with this type of sexual system is dependent on the flow of pollen from the male plants to the female.

[**top**] A close-up of anthurium pollen.

[**bottom**] It's a boy! My *Stephania suberosa* plant (similar to the *Stephania erecta*) only produces male flowers.

Plants in the **Araceae** family (**aroids**) have different flowers and different strategies to prevent self-pollination depending on the genus. Alocasia and philodendron inflorescences contain flowers that are imperfect and unisexual (the tiny individual flowers are either a male *or* a female). However, the problem is, unlike the *Stephania suberosa*, a single plant will contain flowers of BOTH sexes! Their inflorescences will therefore contain a zone of imperfect female flowers and a zone for imperfect male flowers, separated by a sterile male zone, as illustrated in the alocasia inflorescence diagram on page 27. The spathe acts like a bouncer at a club, opening and closing, allowing pollinators to access the different zones at separate times, to avoid self-pollination.

Anthuriums and monsteras, which are also in the Araceae family, are different. They contain inflorescences with bisexual flowers. The strategy of these genera is to go through male and female fertile stages at different times to avoid self-pollination. I'll explain how this works for anthuriums in greater detail on page 117.

POLLINATION AND FERTILIZATION

As you can imagine, if you're going to try your hand at sexual propagation, a thorough understanding of what type of flower you are working with and how it is pollinated in the wild will help you immensely. Scientific publications, books, blogs, and (dare I suggest) social media groups can be valuable sources of information about your specific plant. Often, the closer you can come to replicating the timing and actions of the plant's natural pollinators, the more successful you will be in your pollination efforts. So, let's put on that bumble bee costume and get to work! (Please tag me on social media if you do this).

Sexual propagation can be divided into two big events: pollination and fertilization. Pollination is the movement of pollen from where it is made on one flower to near where the eggs are produced on another flower. Fertilization is when the sperm inside of the pollen unites with the egg inside the flower to form a baby plant. It's important to note that just because you (or that fly, beetle, or bee in the rainforest) purposefully or accidentally pollinated a flower, it doesn't mean fertilization will occur. Just showing up isn't enough, Mr. Pollen!

When it comes to pollinating your houseplant's flowers, what you do and when you do it will depend on many variables, such as what type of plant it is, what type of flowers it has, and its flowers' structure. In chapter 4, I will teach you how to pollinate anthuriums step-by-step because as far as hand-pollination goes, they are beginner friendly. However, let's first begin with a review of the stages of anthurium pollination leading up to fertilization as they occur in the wild.

STAGES OF ANTHURIUM POLLINATION IN THE WILD

Anthurium 1: pollen donor a.k.a. "dad"

Stage One: It's Lady's Night!
Anthurium #1, the "pollen donor," has tiny bisexual flowers covering its inflorescence that first go through a female stage, called **female anthesis**. During this stage, the flowers produce female *stigmatic fluid* (remember, the "sticky stigma" is female). After a few days to a few weeks, it dries up. Time's up, boys!

Stage Two: Male Anthesis
The flowers on Anthurium #1 (the "pollen donor"), then go through their "man era," a.k.a. **male anthesis**. Pollen is produced by the flowers nearest the base of the inflorescence first, and production works its way up the inflorescence over time.

Stage Three: Pollen Removal
A pollinator (typically a fly) removes male pollen that contains two sperm from inflorescence #1.

Stage Four: Pollen Transport
Pollinator flies from Anthurium #1, the "pollen donor," to Anthurium #2, the "seed parent," which currently has flowers in female anthesis. The seed parent is named as such because it will be the bearer of seeds.

Stage Five: Pollen Deposit
Pollinator lands on receptive inflorescence #2 with the pollen on its body. The pollen is deposited on top of the stigma of the flowers, which are all covered in the lady liquid. Okay, fine. Stigmatic fluid. (I just want you to know how hard it is for me to keep my composure here. I'm not mature enough to write this section.)

Stage Six: Pollen Has to Pass Vibe Check
Pollen takes up water and must be recognized by the female flower as compatible (see paragraph about stigma scanning for incompatible plant above).

Stage Seven: Pollen Locks In
Pollen seals the deal, bonds securely to the female stigma, and germinates.

Stage Eight: The TUBE
Pollen grains now grow a pollen tube. Yes, an actual tube emerges from an opening on the surface of the pollen grain. I cannot get over this tube. Humans are all concerned about UFOs, yet pollen is out here growing tubes and knowing where to insert it, and we are acting like there's nothing crazy happening around us.

Stage Nine: Meeting the Bride
The male tube then grows down through the female pistil and into the ovaries of the flower, until it reaches the big lady boss: the ovule!

Stage Ten: Fireworks
The pollen tube then explodes (POW!)

Stage Eleven: Sperm Release
The double-trouble sperm cells are released into the plant ovule.

Step Twelve: Score!
Sperm fertilizes the egg. This happens over and over on each flower of the inflorescence where the pollen successfully germinated.

Mind. Blown.

The end result is a seed that contains three primary parts: the embryo, the endosperm (food storage tissue), and the seed coat. The endosperm feeds the germinating embryo until the seedling can photosynthesize on its own.

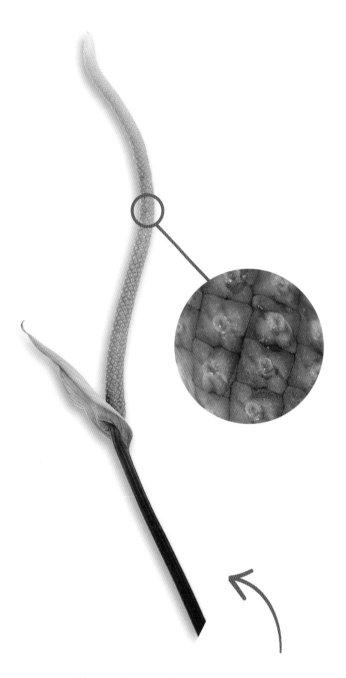

Anthurium 2: seed parent a.k.a "mom" producing stigmatic fluid, and ready for Stage Five: Pollen Deposit.

Tip
The endosperm is why you don't need to add garden fertilizer to seeds for them to sprout. Seeds already have all the food they need to germinate and grow inside of them!

On aroids, such as anthuriums and alocasias, each seed or pair of seeds is encased in a fruit–the berry–that will be an enticing snack for bats and birds. These animals eat the berries when they're ripe and help disperse the seeds after digestion. In other words, when these animals go poo, out come the seeds stripped of the berry goo!

Meanwhile, if Anthurium #1, the pollen donor, also was successfully pollinated (it could have been when it was previously in its female stage and still went on to produce pollen), it's also going through the process of creating berries. If it was not pollinated, it will wither and die.

Cue the Lion King Theme Song: "Circle of Life"

INCENTIVES AND REWARDS

It's one thing for you to purposefully pollinate/sexually propagate your plants with a paintbrush or your fingers at home because you want to make a bunch of free beautiful plants, but what motivates insects to pollinate plants in the wild? Understanding *why* pollinators do what they do in the wild will help you to be more successful as you try, as a human, to achieve what they do. Let's step into the psyche of the plants and the birds and the bees for a moment. (Ideally, you are once again putting on your bumble bee costume as you read this section).

Plants have coevolved with other organisms. For a plant species to survive, they need to reproduce, and in order for reproduction to happen, most plants need to be pollinated. Many plants depend on other organisms for pollination to occur, and we all know how scary it is to have to depend on others for our basic needs. Many flowers, therefore, have rewards to offer pollinators that enter their domain, such as sweet scents, edible nectar, and pollen. And Amazon gift cards.

But how do they attract these pollinators to get their rewards? Since they cannot advertise their floral offerings on social media, plants must be very strategic.

BOTANICAL ADVERTISING

Have you ever wondered why flowers are brightly colored and scented? It's certainly not to attract you at the farmer's market. Colors and scent help to attract various pollinators, such as insects and birds, with the hope that one will transport their pollen to another plant's flower or bring another's pollen to *their* flower. If you start pollinating anthuriums, you'll observe that some flower stalks smell better than others. I once had to move my *Anthurium Besseae* aff. to an entirely differently floor of my house because it smelled like intense mothballs!

Believe it or not, some scents that are harsh to us are highly attractive to certain pollinators. After all, based on my dog walking experience, flies don't have the greatest taste in aromas.

Some flowers, such as those of aroids, even lure in pollinators such as flies, beetles, and bees by producing heat (a process called **floral thermogenesis**). You would need professional tools to be able to detect the temperature difference in your flowers at home, however. Others are known to also release chemicals called volatile organic compounds, or VOCs. Plants are not out here to waste time! They are pulling out all the stops.

ADDED REWARDS

Many plants then reward pollinators who enter their floral chambers with a space to warm up for the night, have insect intercourse, and wine and dine like VIPs in a five-star hotel. On the menu is pollen, floral tissues, and nectar. If pollinators sign up for the plant's reward program, they get extra points towards their next stay too.

If you have a hoya, observe the nectar dripping from the flowers next time it is in bloom. Now you know what that nectar is for! Your plant is rewarding pollinators. The nectar is also super sweet . . . go ahead . . . you know you want to take a tiny taste. Most anthurium stigmatic fluid is also sweet.

Some flowers pull a bait and switch, however: they lure the pollinators in, and once the insects are inside, the flowers physically close around them. Forget rewards, the pollinators are physically trapped inside! After a day or two, once the pollination work is done, the flowers open back up and release them. Tough love.

The next time one of your houseplants is flowering, look at the flower or flower stalk for what it truly is: a flashing botanical motel vacancy sign. A pollinator trap, perhaps. A pollination station!

This examination of both sexual and asexual propagation is necessary for understanding why and how plants are unique in the ways they can reproduce. Whether you decide to make more houseplants through an asexual propagation technique or through pollination–or perhaps both–let's look next at the tools, materials, and environment needed for the job.

FERNS AND SPORES

Have you ever looked on the underside of a fern and seen little brown or black polka dots or thin stripes? They may look funky at first, but this is not a disease. These are called **sporangia**, and they produce small particles called **spores**. In addition to seeds, sexual propagation also includes the reproduction of plants from spores, as is the case with ferns. Fern sex is truly mind blowing and will make you appreciate even more the fact that they have survived this way for over 350 million years—way before the dinosaurs!

This is what survival looks like: First, those brown fuzzy sporangia release thousands of spores on dry days into the air. They are then dispersed by the wind. The hope is that these spores land near other spores from plants of the same species. If a spore lands in a warm, moist place with lots of shade, it will begin to grow into a tiny flat plant the size of a baby's fingernail. This tiny, flat plant contains the organs that produce the eggs and the sperm.

This is where things get good. This tiny, flat plant then goes on with its bad self and produces sperm, which then *swims* a few centimeters in a film of water that covers the plant, trying to find an egg on an unrelated tiny baby fingernail-sized plant. Yes, you read that correctly, the fern sperm swims. Once the sperm finds an egg, they fuse together to create a cell called a zygote. The zygote is then dependent on one of the plants for its nutrition, until the plant has no more to give and dies.

At this point, the young fern is self-supporting with its own root system and fronds. After a few years, the plant will mature and its fronds will create its own sporangia and spores. The fern life cycle is complete! There are collectors of ferns who propagate at home, but as you can imagine, this is not for the beginner. They collect spores from the sporangia when ripe onto a piece of paper and then plant them into a damp sterile potting mix, keep it covered, and hope the above process will take place. I have never tried this myself, but it sounds like a fun activity to do spor-atically!

I once thought this was a nasty fungus on my staghorn fern, but the tiny brown specks are actually spores.

Chapter 3

TOOLS, MATERIALS, AND ENVIRONMENT

Many types of asexual plant propagation are a race against stem and root rot. The race starts as soon as you take a fresh cutting of plant tissue to propagate. If you win, your cutting produces new growth before any rot spreads to the areas where new roots and shoots grow from. If a fungus or bacteria wins, rot spreads to the critical areas of new root and shoot growth, and the cutting dies.

A layer of bacterial or fungal rot can occur on plant cuttings in any substrate, often starting on the bottom tip of the cutting as a black or dark brown spot. A little bit of rot can always be cut off with a clean pair of plant snips. Oftentimes, once a cutting begins to grow fresh roots, the rot will cease to spread altogether.

However, if the rot begins to spread upward and you forget to cut it off in time, it will become a problem. Rot is caused by bacteria or fungi that grow and spread in damp conditions, especially those that lack significant amounts of available oxygen (otherwise known as **anaerobic** conditions).

A cutting rooting in water that is left unchanged for prolonged periods of time is at higher risk for rot because water loses oxygen over time. Oversaturated soils also cause oxygen levels to drop significantly, leading to an increased risk of rot. Chunky potting mixes allow for more air pockets between particles and create a more **aerobic**, oxygen-rich environment. Rot also can be caused by fungal infections introduced to the plant by contaminated pots and tools.

Expect to lose the race from time to time. Every plant parent does, especially if your environmental conditions are not ideal. But now that you have this book in your hands, your chances of winning are already looking much better. And, if you continue to read the book, the odds will forever be in your favor! The key is to get your cuttings to root quickly while preventing pathogens from entering the wound site and spreading.

In this chapter, you'll learn how to win the race nearly every time through the optimal use of:

1. Tools
2. Substrates
3. Environmental Conditions: including light, humidity, and warmth
4. Rooting Agents
5. Oxygen

This race was a close one, as rot was spreading quickly up my variegated *Monstera deliciosa* 'Aurea' and consumed the bottom node. However, just in the knick of time, a white aerial root popped out of the second node on top. Now I can hopefully chop off the rot below the new root and it will continue to develop. If I had just let this cutting be, there's a good chance the rot would have consumed the entire cutting. Always monitor cuttings closely during propagation.

PROPAGATION TOOLS

The tools you'll need to propagate your plants are somewhat dependent on what types of plants you're propagating. Let's look at tools for propagating the two most common groups of houseplants: tropical plants and succulents.

TROPICAL PLANT PROPAGATION TOOLS

Here's the truth: you don't *need* many tools to propagate plants. However, it's definitely fun to have a bunch of cute little tools on hand to use. The only tool you truly need for most tropical houseplant propagation projects is something sharp and sterilized to cut plants with. This could be a pair of small **gardening shears**, scissors, an X-acto knife, or your clean sharp fingernails (savage). I own no less than five pairs of small, sharp gardening shears and use them for everything from culling dead leaves to taking stem cuttings to trimming my son's hair during the pandemic lockdown (true story).

Additionally, it's a good idea to have a **magnifying glass** on hand to inspect for pests while you're all up in your plant, cutting it to bits. I recommend buying a 45x magnifying glass with a light like the one you see in the photo. Alternatively, you can use a jeweler's loupe or, if you're a tech person, one of those handheld microscopes that connect to your phone or computer.

If you plan on growing tropical plants from seed, you may want to have a **paintbrush** for removing and transferring pollen and a **seed dibber** to poke small holes in potting medium to plant seeds in. The holes prevent seeds from floating around during watering. I don't find either of these tools to be absolutely necessary when you have ten fingers at your disposal, but having them certainly makes the jobs of pollination and seed sowing less messy.

45x magnifying glass with light

Seed dibber

Paintbrush

Gardening shears

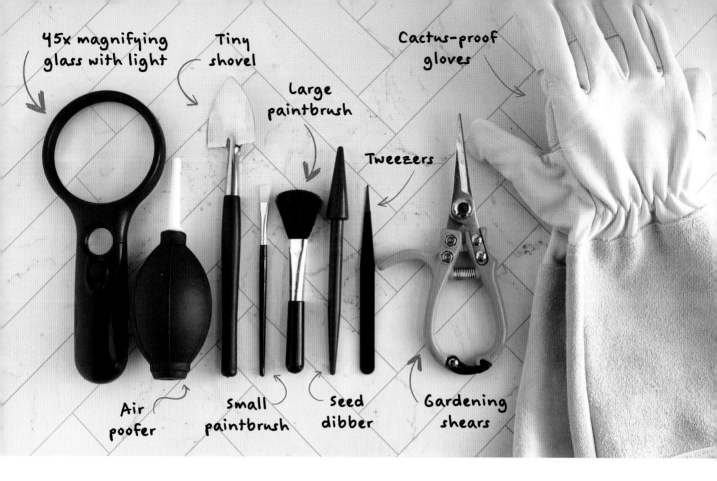

45x magnifying glass with light

Tiny shovel

Cactus-proof gloves

Large paintbrush

Tweezers

Air poofer

Small paintbrush

Seed dibber

Gardening shears

SUCCULENT PROPAGATION TOOLS

A simple search for "succulent propagation tools" online reveals several cute tool kits and supplies for you to purchase. These tools are important for use with succulents in particular because of their compact, delicate, and sometimes prickly nature. Unlike with tropicals, there are many times when your fingers may be too thick for planting dainty little succulent arrangements, so using a thin planting tool will behoove you.

Just as with tropical plants, you'll want to have a **magnifying glass** on hand to search for pests. On succulents you'll want to look out for white, fuzzy mealybugs especially. Also, as with tropical plant propagation, you'll want to have a pair of sterilized **cutting shears** and, if planting seeds, a **seed dibber** (very optional).

Because succulents often have spines or can be rosette shaped, it's not uncommon for dirt and debris to get lodged between plant parts. An **air poofer** will help clean your plant by dislodging stuck potting mix and dust after repotting or shipping. Learn from my experience—using an air poofer is better than blowing air out of your mouth to clean your plant, which results in getting potting mix directly in your eyes. **Brushes** and **tweezers** also help with cleaning. Tweezers also are helpful if you are planting tiny terrariums.

Since I was a jerk and tried to plant a cactus without gloves during the making of this book, I needed said tweezers to pull **glochids**—sharp cactus bristles—out of my fingers as well, although I know this is not why they include them in succulent tool kits. That brings us to our most critical spiny cactus-handling tool: **thick, protective gloves** to avoid having using your tweezers for their unintended purpose. **Tongs** and **paper bags** (not pictured) also help to pick up spiny cactus pads.

Last but not least, my favorite tool—the **tiny shovel**. Not only does it look ridiculously cute, but it's helpful when potting tiny succulents into tiny pots. Cute, cute, cute.

TOOL STERILIZATION

No matter what type of plant you are working with, it's a good practice to sterilize your cutting shears prior to pruning or propagation, or at least make a habit of cleaning them after using them. Does everyone always do it? No. But it's good practice in case there are lingering pathogens (bacteria/fungi) on the blades that could enter the tissues of the next plant you use them on.

Sterilization becomes absolutely necessary when working with a clearly diseased or pest-infested plant. In this situation, you'll want to sterilize your cutting shears between each cut you make to avoid pathogen spread. It wasn't until I started growing anthuriums, in which bacterial and fungal infections are a concern, that I realized the strong importance of cutting tool sterilization. Prior, my propagation and pruning experience was limited mainly to hardier houseplants such as hoyas and philodendrons. There's a reason why houseplants such as a Christmas cactus or a snake plant or a ZZ plant are so popular; they are easier to grow and are more resistant to pests and disease than sensitive velvety anthuriums or delicate collectors' begonias. When working with species more susceptible to pests and disease, sterilization is a nonnegotiable.

How to Sterilize Cutting Tools

You can sterilize your cutting tools in a number of ways to kill off pathogens and pests, but perhaps the easiest and most accessible is to use 70 to 90 percent rubbing alcohol. You can buy it already in a spray bottle from the drugstore or just buy a bottle and pour it into a spray bottle yourself.

To sterilize, first wipe away all plant material from the blades of your cutting tool and then spray it with rubbing alcohol. The rubbing alcohol will evaporate quickly, so simply spray it and allow it to dry on its own without wiping it off, or wipe it off after a few minutes. Other effective disinfectants include bleach, which I don't recommend unless you love stained clothes, and any quaternary ammonium disinfectant such as Physan 20, which is popular in commercial greenhouses.

Wiping down my cutting shears with rubbing alcohol like all the cool, clean kids do.

You will be most successful propagating plants in potting soil when working with hardy succulents such as snake plants. This is not to say you won't be successful with sensitive tropical plant cuttings placed directly in potting soil straight from the bag; you'll just have to monitor them more closely.

SUBSTRATES

Why use a specific propagation potting substrate (a.k.a. growing medium), you ask? Why not just plop all your cuttings in whatever general potting soil you have on hand? Or, better yet, why not simply cut up your plant and stick it back in the pot with the mother plant and see what happens? After all, this is the path of least resistance.

I never discourage experimentation, and depending on what type of plant you're working with and your environmental conditions, you may or may not be successful in getting plant babies to root in regular potting soil. While simply planting your propagules straight into potting soil is easy, it does not always have the highest rate of success, especially when working with sensitive tropicals.

When "straight to soil" works during tropical plant propagation, it produces robust plants quickly due to the nutrients in the potting soil, but the watering frequency and humidity must be just right. Usually, however, tropical plant leaf or stem cuttings require higher humidity and more consistent moisture than the mother plants are living in to take root, so putting the cuttings back in the pot with the mother plant doesn't always provide them with the

proper environmental conditions for success. Other times, when planting propagules directly into potting soil that remains moist the entire time, one will lose the race: the rot will creep up the bottom of the cutting and into the node before the roots establish themselves. This is because potting soil is often too dense for propagation, creating anaerobic conditions (not enough oxygen). Plus, some types of potting soil are not sterile (more information about that in the next section).

This all being said, propagating directly into potting soil works more often with succulents. Because you are watering the mother plant so infrequently to begin with, and no extra humidity is needed, there is little opportunity for the leaf or stem propagule to rot.

PROPAGATION SUBSTRATE REQUIREMENTS

I know from my own experiments that which substrate you use for propagation makes a huge difference in how fast your propagations grow. There are **organic substrates**, like coco peat and peat moss, and **inorganic substrates**, such as pumice and perlite, and with so many of them on the market, it can feel overwhelming. There are almost as many substrates in which to propagate your plant as there are ways to cook an egg.

To simplify, an ideal substrate for propagation should fit two main requirements:

1. **Well-draining.** *Is the substrate coarse enough? Does it contain a lot of air pockets between individual pieces/granules/fibers to provide space for oxygen?*

2. **Moisture retaining.** *Can it hold moisture like a sponge? If it feels hard like a rock, is it porous and absorbent?*

A third factor, which is not a requirement but an added bonus during the sensitive propagation phase, is that the medium is sterile. A sterile substrate does not contain a high concentration of living microscopic organisms, such as bacteria, that can lead to root rot. To win the race against rot, some people choose to propagate their houseplants in a sterile organic medium, such as dried New Zealand sphagnum moss or inorganic perlite. The sterilization processes these substrates go through after harvesting kills the pathogens that may contribute to root rot.

Tip

Organic substrates are derived from living or once-living materials such as plants. Inorganic substrates are made of nonliving materials such as rocks and minerals.

I love these little figurines that are designed to hold cuttings up straight in a glass of water.

The following few sections review the most common substrates used for propagation and as components of potting mixes. Please note that not pictured in this section are pumice, sand, compost, and biochar. It's not that I don't believe they are important, but they are not substrates I regularly rely on.

PROPAGATING IN WATER

The most popular propagation medium is old faithful H_2O. Propagating in a glass of water is quick and simple. After the cutting has grown branched roots that are 1½ to 2 inches (4 to 5 cm) long (or longer than a person would like to admit if they've forgotten about their propagation project), they transfer it to a potting mix or an inorganic medium of their choice, such as LECA or pon (see page 51). It is interesting to consider that water is such a useful propagation medium, considering it isn't sterile or chunky. At least I hope your tap water is not chunky!

Propagating plants in water has a few key advantages. The first may be obvious, but water keeps your propagules hydrated and takes the guesswork out of watering! It's also readily accessible. Many people use distilled water for propagation, but tap should be fine in most cases, depending on the quality of your water. If you do use tap water, it will save you money versus other more expensive substrates.

Because water is transparent, it is easy to monitor your propagules for stem and root rot. This makes water a popular choice for sensitive variegated plants that are highly susceptible to rot, or for cuttings that already have some rot and you want to monitor it daily. Transparency also allows for monitoring root growth.

Water propagation is not all rainbows and unicorns, however. It is also green, slimy algae. Algae forms when bright light hits areas of moisture for prolonged periods of time, so you need to change out the water to keep it fresh or use a dark-colored glass vessel.

Though there is some dissolved oxygen in well-aerated water when it first comes out of the faucet, it is not a never-ending source of oxygen. Plants need to uptake oxygen molecules for **respiration**. This oxygen is necessary to breakdown the sugars produced through photosynthesis. The energy produced from this process is then used to power different functions throughout the plant such as the growth of new roots. Over time, when rooting a cutting in water, the oxygen supply becomes depleted, so one must either change the water frequently or use an aquarium pump and air stone to create an aerobic environment.

Finally, water propagation can sometimes take longer than propagation in other mediums or methods, and roots that grow in water may need a little extra TLC when transitioning to a solid medium. Water is an amazing medium, but it's definitely not all the be-all and end-all! In the next section, we will discuss some of your other options for propagation.

CONVENTIONAL PROPAGATION SUBSTRATES

The following are substrates and components of conventional potting mixes that are used for traditional houseplant growing where you would typically water your plant from above (as opposed to a semi-hydroponic or hydroponic setup).

[A] PEAT MOSS: Mainly decomposed once-living sphagnum moss · Essentially sterile and contains very few nutrients · Acidic pH · Harvested from wetland peat bogs · Once considered the coolest soil-amendment in town for the way it retains moisture and lightens up soil, it's now viewed as the environmental mean-girl · Sales in Europe to gardeners were banned starting in 2024 · Peat takes thousands of years to regenerate · Peatlands are our biggest terrestrial carbon store and harvesting peat releases this carbon, which contributes to global warming · Harvesting destroys an essential habitat for some incredibly rare species of plants and animals · Compost and coco coir are the most viable alternatives

[B] COCO COIR: Outer shell of a coconut all ground up · Also called coconut coir or coco peat · Coarse fibers and dust · Neutral pH · Slightly more sustainable than peat moss · There are environmental concerns over chemicals needed to process it, waste products produced as a result of processing, fossil fuel costs to transport it, and concerns over worker safety and their respiratory health

[C] COCO COIR CHIPS: Chips made from the outer shells of coconuts · Spongey chunks retain a lot of moisture and aerate the soil · Similar to coco husk

[D] PINE BARK: Often sold as "Orchid Bark" · Wood chips from the conifer *Pinus radiata*, also known as the Monterey pine, in New Zealand · Aerates the soil · Does not break down easily · Absorbs moisture · Great for epiphytic and semi-ephiphytic plants that grow anchored to tree trunks (like orchids and aroids) · Fir bark is a simliar and commonly used alternative for orchids

[E] VERMICULITE: Flaky, shiny, mineral particles that absorb water and aerate growing media · Heating process during production makes it more or less sterile · Has some potassium, magnesium, and calcium · Neutral pH

[F] EARTHWORM CASTINGS: Earthworm poopies! • Also known as vermicast • Organic fertilizer: contains macro and micronutrients • Can be sprinkled on top of houseplant soil or mixed into a potting mix • Will not burn plants

[G] PERLITE: Crunchy, expanded, volcanic glass bits that absorb moisture • Creates air pockets in potting soil • Relatively low cost • Can be purchased in different sizes from fine to coarse • Used in convential growing and semi-hydroponics • No essential plant nutrients • Neutral pH

[H] ACTIVATED/HORTICULTURAL CHARCOAL: Obtained by the burning of wood waste and coconut shells • Slightly pourous • Absorbs some moisture • Absorbs fertilizer and slowly releases it back into the soil • Assists in soil aeration • Reduces soil acidity • Helps eliminate bacteria and odors

[I] TREE FERN FIBER: Derived from the trunk of tropical plants called tree ferns • Tree fern fiber from New Zealand is a good choice because it's considered sustainable • Excellent drainage, retains water, aerates soil, and long lasting • Slightly acidic pH makes it ideal for anthuriums, orchids, and ferns

[J] SPHAGNUM MOSS: Dried moss often harvested from wetlands in New Zealand and other parts of the world • Incredible moisture-retaining properties • Acidic pH • Used often for propagation or as a soil ammendment • Used often by carnivorous plant and orchid growers and houseplant propagators • Don't accidentally drop a strand on your floor; you'll later think it's a centipede and freak out

I grow the majority of my alocasias in pon (see page 51) in self-watering pots. The meter on the side of the pot tells me how much water is left in the reservoir on the bottom of the pot and when the reservoir has reached max capacity.

INORGANIC SUBSTRATES AND SEMI-HYDROPONIC PROPAGATION

When growing and propagating houseplants, the most common watering method is to actively pour water onto the substrate from above. This can take a lot of time and effort if you're watering up to twenty, thirty, or three hundred plants and propagations. Maintaining a steady watering schedule also can be a challenge for people who travel a lot. Finally, plants have different watering needs throughout the year, and it can be difficult to guess at what time intervals to water, especially for delicate, sensitive plants.

Therefore, an increasingly popular alternative to conventional growing and propagation is **passive-hydroponic** or **semi-hydroponic** growing and propagation. In semi-hydroponic growing, the plant's roots obtain the water and nutrients they need by absorbing it from a reservoir on the bottom of the pot, soaking it up through one pebble of substrate at a time. This type of absorption is called **capillary action**. Ideally, the plant's roots do not physically touch the reservoir of water on the bottom, only the substrate particles.

Have you seen advertisements for self-watering pots? They are designed to make your life easier, so you don't have to water as frequently. Self-watering pots utilize a semi-hydroponic method of growing.

Houseplant growers and propagators love semi-hydroponics because it takes the guesswork out of watering. The plant absorbs the amount of water it needs, as it needs it, as long as you keep the reservoir of water on the bottom filled. Since you aren't watering on a regular schedule and you're only refilling the reservoir once it runs empty, you can leave your plants and propagules for a long period of time while you're on vacation without worry.

Instead of using a chunky potting soil mix, the semi-hydroponic propagation method typically uses what looks like a bunch of inorganic pebbles or rocks. The most popular inorganic substrates used in semi-hydroponics, as shown to the right, are perlite, LECA, and a mix called pon.

INORGANIC SEMI-HYDROPONIC SUBSTRATES

Perlite

Puffy porous bits of volcanic glass •
Used frequently in conventional
potting mixes as well

LECA

LECA stands for Lightweight Expanded Clay
Aggregate • The cocoa puff dupes that your
plants will go cuckoo for • Porous baked
clay balls • Does not contain nutrients so a
nutrient solution must be utilized with it

Pon

A mixture of pumice, lava rock, and zeolite • Made
famous by the company The Horst Brandstätter Group
that sells it under the brand name Lechuza • You can
DIY pon by purchasing the ingredients and mixing on
your own • Personally, I love the convenience of buying
Lechuza pon (when it is available) because I know I
am getting the same quality and ratios every time

In addition to the substrates mentioned above, less commonly used semi-hydroponic substrates include rockwool and growstones. All these substrates are sterile and therefore contain fewer pathogens and microbes that can potentially lead to root rot. Additionally, because of using them, you also will have fewer soil-dwelling pests, like fungus gnats—although I must warn you that I recently I had a fungus gnat infestation in pon (those a-holes will lay larvae anywhere when they're desperate enough).

While propagating in a semi-hydroponic setup is very easy and low maintenance, if you plan to grow your plants in such a system for a long period of time, it requires periodic maintenance to keep the pots, substrates, and reservoirs clean, and the substrates themselves can be expensive. Most people flush excess salts from the substrates at least twice a year. This is not as relevant for propagation as it is for growing because the propagules are not going to stay in the propagation substrate long enough

Most plants root extremely well in Fluval Stratum. Simply take a healthy stem cutting and place it in a plastic cup filled with Fluval Stratum. Fill the cup roughly ¼ to ⅓ of the way up with water when the Fluval looks dry. You'll know when Stratum is dry because the Stratum will appear lighter in color.

to warrant a flushing before the roots have grown and the plant is ready to transfer to a new pot.

Finally, besides the temporary slow-release fertilizer included in the original Lechuza-pon substrate, these inorganic substances don't contain nutrients, so you need to add a nutrient solution to the water when growing mature plants in a semi-hydroponic setup. When propagating, however, since it's short term, a nutrient solution isn't necessary and may actually slow the rooting process. Cuttings don't uptake nutrients which must be taken up by roots, so until roots are present, fertilizer doesn't do any good. Adding fertilizer to unrooted cuttings may even encourage the growth of the bacteria that cause rot.

FLUVAL STRATUM

Fluval Stratum is becoming increasingly popular for houseplant propagation, and it is so unique that I wanted to dedicate an entire section to it. Fluval Stratum, also known as shrimp Stratum, has been used for years in the aquarium hobby in planted tanks and for people who grow freshwater shrimp.

Fluval Stratum may look like tiny, inorganic rocks, but don't let it fool you. It is made of organic, compressed volcanic soil! And what blows my mind about Stratum is that this soil comes from a single source: the foothills of the Mount Aso volcano in Japan, where the soil is rich in nutrients and minerals. Field trip anyone?

Not only does Fluval Stratum *provide* nutrients, but it is the ideal pH for plant nutrient uptake. It even can slightly lower the pH of the water surrounding it as well, often to an ideal pH for most houseplants, 6 to 6.5. When the water pH is in the ideal range, plants can better absorb nutrients. This is why propagations take off in this magic volcanic caviar as soon as they develop roots.

There is no need to rinse Stratum before use, and there are no strict watering guidelines. Some people use it conventionally and others semi-hydroponically. I utilize it for propagation by first filling up a small plastic cup with Stratum and then sticking a propagule half of the way down into the substrate. I then add water, so it fills up approximately ¼ to ⅓ of the cup. When the Stratum starts to dry out completely, it becomes lighter in color. When there is no water left on the bottom of the cup and the Stratum is mostly dry, I simply pour more water into the top of the cup so that the bottom ¼ of the cup is filled again.

It's okay if you add more or less water. Experiment with it. Stratum is forgiving!

The best part of using Stratum is not the ease of use, its nutrient supply, or its pH. It's the ASMR sensory experience that it provides. Stratum is the Rice Krispies of the plant world. If you put your ear up to it after you pour the water in, you can hear Snap, Crackle, and Pop. Try it!

One drawback to Stratum is that it is delicate and will degrade over time. Therefore, it's recommended that you replace it every two years. While it is easily accessible and you can buy it at most major pet stores (where it's being sold for aquariums), it is expensive. Therefore, you can also mix it 50/50 with perlite to make your supply go farther.

The other big disadvantage is that it is literally the worst possible substrate to spill. If you try to wipe it up, the soft granules immediately crush into the carpet or whatever surface it landed on, creating a powdery, charcoal-like mess. Herein lies the need for an important disclaimer:

All the Plant Babies, LLC cannot be legally responsible for any emotional damage that arises from the inevitable spilling of Fluval Stratum used as a direct result of you reading this book, purchasing some, and then using it for propagation.

[below] I planted *Oxalis triangularis* corms in Fluval Stratum two months ago, without sprouts or roots . . . and look at them now! The glass has no drainage hole, and I fill the glass with roughly ¼ to ½ inch (6 to 13 mm) of water every week. The plant receives bright, indirect light.

CHOOSING A PROPAGATION SUBSTRATE

You know what plant you want to make more of, and (now) you know more than you ever wanted to know about all the available substrates. But how do you know which substrate to use on the specific plant you're propagating? Overall, if you understand your plant's environmental and watering needs, no matter what you decide to prop in, your success there is no stoppin'!

The substrate you choose depends greatly on your level of experience with growing and propagating plants indoors. Water is the best medium for beginners and the easiest to work with. If you took some cuttings and you're a play-it-safe kind of plant parent, plop them in a glass of water and be gone with your bad self.

On the other end of the spectrum are the tinkerers and the thrill seekers. While plant people aren't exactly known for being adrenaline junkies, when it comes to getting fancy with propagation substrates, some of you are just W I L D. Recently, for example, houseplant collectors in an online group I'm in are having success mixing perlite and Fluval Stratum together with a layer of LECA on the bottom (a semi-hydroponic setup). I've seen people mix several crazy concoctions together and often have excellent results. This crowd tends to be a bit more experienced with houseplants.

To help guide you through the substrate-selection process of someone who propagates plants for a living (me!), I've created the following chart, which outlines five different theoretical propagation scenarios and how I would approach each one.

THE PROPAGATION-SITUATION (MY MTV *JERSEY SHORE* NICKNAME)	IF YOU PLAN TO EVENTUALLY POT YOUR PLANT IN . . .	CONSIDER PROPAGATING IN . . .	WHY? BECAUSE:
You're new to houseplant propagation and you want to root cuttings of your easygoing pothos.	Houseplant potting mix out of the bag (and perhaps add a handful of extra perlite for more drainage if you're feeling fancy)	Water!	It's easy and will work just fine for a no-fuss plant, such as a pothos or trailing heartleaf philodendron. See page 80 for instructions.
You want to propagate your string of hearts using the butterfly method.	Succulent potting mix	Succulent potting mix	While there may be a greater risk of rot propagating "direct to soil" as opposed to say, sphagnum moss first, the benefits are that it produces faster results and more robust growth. See page 164 for instructions.
You have indoor tropical plant-growing experience and want to propagate your notoriously difficult -to-propagate variegated monstera.	Aroid mix	Water, then sphagnum moss	Root first in water so you can monitor carefully for root rot, changing the water frequently and cutting off rot if you need to. Once roots are at least 1½ inches (4 cm) long, transition to moist sphagnum moss and allow roots to grow another inch or so. This will enable a smooth transition from water to potting mix, and there will be less of a chance of the monstera crashing.
You are a hoya enthusiast and want to propagate your hoyas quickly.	Pon or LECA in a self-watering pot	Fluval Stratum and/or perlite	Fluval Stratum is made up of small particles, which are beneficial for tiny roots and will acclimate plants to a semi-hydro setup from the start. Fluval Stratum is widely available from pet shops and online. Perlite also is widely available and also can be used in the same way. Many people mix them together 50/50 to save money because Stratum is expensive.
You are new to propagating succulents.	Succulent potting mix	Literally nothing. Just let them lay out on a shelf.	Succulents are crazy. I don't make up the rules. See page 142 for more on propagating succulents.

Not mentioned in the chart on the facing page is the process for selecting a substrate for seed germination (sexual propagation). While the decision-making process will be somewhat similar, growing plants from seed is a completely different process than growing roots on vegetative cuttings. I will cover how to grow anthuriums from seed in chapter 4 in further detail.

SOILS AND POTTING MIXES

Are you sitting? Okay, good. This may be big news: Your houseplant potting soil may not contain any actual "soil."

I know. Take deep breaths. This will take some time to digest.

Most people grow up thinking soil is soil and that all plants simply grow in the same brown stuff. Or, they grew up without giving much thought to what plants grow in at all (gasp)! However, outdoor soil (also known as native soil), garden soil (also known as topsoil), indoor potting soil, and potting mix are completely different things. Oy vey.

If this sounds confusing, it's because it IS! Never fear though. Throughout this next section, I am going to simplify all of this for you, and spill the dirt, on dirt. While propagation substrates are our main concern throughout this book, it's important to know a bit of background information on what soil and potting mix is before you start amending them to suit your propagation needs.

The following is a summary of the soils and mixes you will hear about most often as a houseplant propagator:

- **Natural soil** is the combination of mineral and organic materials on the surface of the Earth where plants grow. It is also referred to as native soil, as I mentioned previously. Look out your window and onto the ground nearby. Can you spot a patch of native soil? You will NOT be using this for your houseplants or propagations.

- **Garden soil**, a.k.a. topsoil, is, you guessed it, what you use outside in your garden. You buy it most often in a big bag at a garden center or at a big box store. Garden soil is a mixture of screened native soil, minerals, and organic matter. It retains a lot of water and is mixed into garden beds to improve the native soil composition and make it easier to grow outdoor plants. Depending on the brand, garden soil may be sterile and pathogen-free, or it may contain pathogens, fungi, and bacteria because it is not sterile. Regardless, you will NOT be using this for your houseplants or propagations.

Pretty and gritty: Cacti and succulent potting soil

- **Potting soil or potting mix** is the type of "soil" you want to use on your houseplants and propagations (even though it doesn't include any true soil at all). The terms "potting soil" and "potting mix" are used interchangeably. These products do not contain any true soil; they are a sterile growing media (unless compost or another organic matter is added to the product to introduce beneficial microbes). Potting soils and mixes often contain the same ingredients but in different ratios, depending on the brand. The primary ingredient in potting soil is often peat moss (a.k.a. sphagnum peat moss—not to be confused with the spongey, stringy, long-fiber sphagnum moss) or coco coir. It usually also contains perlite or vermiculite along with lime to balance out the pH and a wetting agent to help it retain moisture. Some brands contain additional ingredients such as pine bark or rice hulls. Organic or synthetic fertilizers may also be mixed in.

- **Seed starting mix** is light with fine particles. When it comes to seed starting substrates, opt for lighter and finer, not nutrient dense or chunky. Light and fine means it won't contain heavy chunks that delicate tiny roots have trouble growing around. These mixes are made typically of peat moss or coco coir and perlite and/or vermiculite and lack fertilizer since seeds contain all of the nutrients they need for the very beginning of life inside their coats. Sometimes, however, seed starting mixes include a little bit of worm castings, as some believe these help speed up seed germination. When planting aroid seeds, my experience and multiple experiments have shown that seeds germinate and grow a lot faster in substrates such as Fluval Stratum or sphagnum moss than in a traditional peat-based seed starting mix.

- **Propagation mix** is a combination of substrates blended together to root propagules in. Alternatively, you can use any of the substrates listed in the pink box below by themselves for houseplant propagation, without mixing them together. By blending them, however, you can harness the benefits of multiple substrates or soften the financial blow when it's cost prohibitive to use one of the more expensive ones in large amounts (I'm looking at you, Fluval Stratum). Personally, I am a big fan of mixing any substrate with perlite, as it is sterile, airy, absorbent, and low-cost. For these reasons, it's probably the most commonly used ingredient across various propagation mixes.

Propagation Mixes *

Perlite + peat moss or coco peat (50/50)
Perlite + sphagnum moss (30/70)
Perlite + Fluval Stratum (50/50)
Coarse perlite + LECA (50/50)
Coarse perlite + LECA + vermiculite (40/40/20)

*Note: These ratios are merely suggestions based on my personal experience. Make your mixes based on what you have available and your own preferences. Never stop experimenting!

- **Aroid mix** became popular over the past few years alongside the rise in popularity of aroids such as philodendrons and monsteras. Aroid mix is exactly as it sounds: a potting mix that is tailored for the specific needs of aroids. As most aroids are either **epiphytic** or **semi-ephytic**, aroid mix includes ingredients that are moisture retaining and yet bulky and therefore dry out quickly. Epiphytic plants typically grow on trees in the jungle and obtain moisture and nutrients from the rain and mist in the air. Semi-ephytic plants can grow either as an epiphyte or rooted in the soil. This tree bark and/or coco chip-amended substrate creates a root zone environment that most closely mimics what aroids thrive in naturally.

Aroid mix also is used for other plants, such as hoyas and ficuses. Aroid mix can be amended to make it more or less chunky, depending on whether you are growing plants that are epiphytic, semi-ephiphytic, or **terrestrial**. Terrestrial plants are those that grow straight in the soil in the wild. Some people amend their aroid mix to be less or more moisture retaining, depending on if they tend to overwater or underwater their plants.

The photograph on the next page illustrates what I use for my aroid mix. You can also purchase premade bags of aroid mix from specialty stores online. But while convenient, it's more expensive than making your own. Additionally, you can break this recipe down even further. Instead of using indoor potting soil, use peat moss, earthworm castings, crab meal, bat guano, and other additives. Just know that if you do make your own mix completely from scratch, it will lack a wetting agent to retain moisture like most indoor potting soils include, and most importantly, unless you add it, the mix will not include the proper amount of lime, which potting soil often includes to perfect the pH of the mix. For this reason, I choose not to make my own 100 percent from scratch and instead I use this recipe:

Aroid Mix

50% High quality organic potting soil

20% Pine bark/ orchid bark

20% Perlite

10% Horticultural charcoal *

*Charcoal is a nice-to-have, but not a make-or-break ingredient. If you run out of it or can't obtain it, don't worry. Simply substitute with extra pine bark and/or perlite.

- **Anthurium mix** is created for growing epiphytic and semi-ephytic anthuriums that is usually more coarse and airy than your average aroid mix.

That being said, I've seen them grown successfully in everything from thick, straight-up peat moss in Florida nurseries to a chunky coco husk-based mix with ten ingredients. When growing conventionally, there is no one mix that works for everyone.

If you are just starting out and growing the conventional way (as opposed to a semi-hydroponic setup), I recommend starting off with a mix that is moisture retaining, has a lot of air pockets, and dries out quickly. One option is to pot your anthurium up into something along the lines of:

> 60% Aroid mix (as described previously)
> 20% Tree fern fiber
> 20% Additional coco chips

Over time however, my personal mix has evolved to look like this:

> 20% Coarse perlite
> 20% Pine bark
> 20% Coco chips
> 10% Tree fern fiber
> 10% LECA
> 10% High quality organic potting soil
> 10% Charcoal

With this mix, I water the seedlings every 5 days, but I leave a lot of water in the trays so that they can soak up the water from the bottom, like a reservoir, over the 5-day period and never dry out. I also give my anthuriums a flushing every few waterings. They are in 60 to 80 percent humidity.

Another popular anthurium mix that people use at various ratios include:

> Coco husk
> Coarse perlite
> Pine bark
> Tree fern fiber or LECA
> Charcoal

As you can see, you have many options. Experiment with your seedling mix and make it chunky, but never let it dry out before your next watering!

Cacti and Succulent Soil Mix

This mix is suitable for plants that retain a lot of moisture in their leaves. Succulents enjoy a gritty, well-draining mix. Succulent soil mixes typically contain amendments such as sand, gravel, and perlite.

For my succulents, I'll be the first to admit (and you won't be surprised if you read the anthurium mix section), I get a little bougie. You can make a much simpler and less expensive mix with any number of the recipes out there. However, I've been using this combination for years. It works well to prevent root rot because it's so airy:

10 to 20% perlite or pumice

60% store-bought cacti and succulent mix (straight out of the bag). Ingredients: compost, peat moss, sand, limestone (pH adjuster)

20 to 30% specialty gritty succulent mix (straight out of the bag). Ingredients: fired clay, montmorillonite clay, pine coir

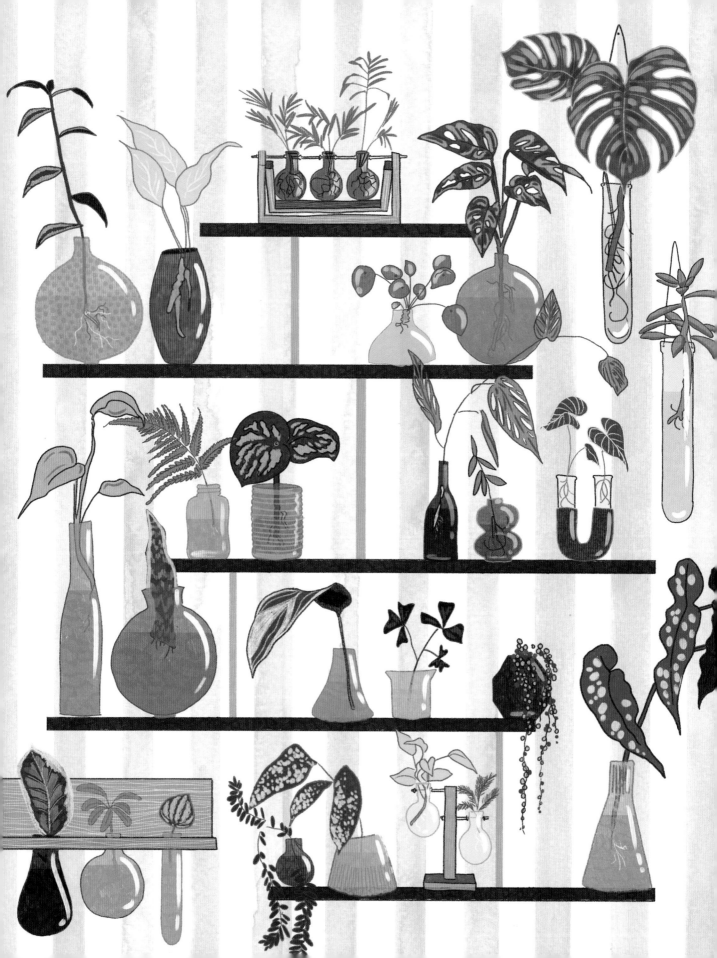

TRANSITIONING ROOTED PROPS OUT OF WATER

Many people do awesome with water propagation, as will you after reading this book. But then, after the plants have roots, they transfer them to potting mix and the plant rots and dies. It has happened to me. I'd rate that experience a 0/10. Would not recommend.

This common propagation tragedy has to do with the fact that while the roots a plant grows in water are the same roots that would grow in any other substrate, they are physically different than those that develop in soil. Within the houseplant hobby community, people refer to roots that develop in water as "water roots." Water roots are clean, smooth, and less hairy, like your own body after the day at the spa (although water roots do sometimes develop green algae and I hope that does not happen to you at the spa. Roots that develop in potting soil/mix, a.k.a. "soil roots," tend to have longer root hairs.

While the difference between the roots that grow in water and those that grow in a potting mix may not significantly impact hardier plants like pothos, it can cause a sensitive plant like a *Monstera deliciosa* 'Albo-variegata' to crash and die. Therefore, to ensure a smooth transition from your propagation medium to your final medium, I suggest the following:

1. When transitioning from water to potting mix, ensure your potting mix is extra airy and coarse. Add extra orchid bark and/or perlite to that bag of store-bought potting mix to ensure it's light and water drains through it quickly. If your potting mix looks like thick brown mud when wet and water pools on top when you water it, it is NOT airy and fast draining and does *not* have a lot of oxygen in it.

2. For the first 1 to 2 weeks after the transition, keep your plant in as high of humidity as you are able to, then gradually ease it into its permanent new environment.

3. Consider propagating from the start in a solid substrate, such as sphagnum moss, perlite, LECA, or Fluval Stratum (or a mix). This way, the plant's roots develop already adjusted to a solid medium.

4. For very sensitive rot-prone plants like variegated monsteras, consider an in-between transition step. After water propagation, allow them to root in, and adapt to, a VERY airy and sterile solid substrate that's continuously wet for a few weeks, such as LECA or perlite or sphagnum moss, and then move it to potting mix. (The reason why I root variegated monsteras in water in the first place is because of the ability to monitor for rot daily. Using an aquarium pump with an air stone in the water is a bonus).

ENVIRONMENTAL CONDITIONS: LIGHT, HUMIDITY, AND WARMTH

You may want to propagate some of your houseplants but you're wondering: Is the timing right? To which I will answer with a quote from one of my favorite spiritual teachers, Eckhart Tolle, "The present moment is all you will ever have." (I know he was not speaking about houseplant propagation but roll with it).

Unlike outdoor gardening, where conditions are a little more out of your control, indoors, you can control the environment to meet your plants' needs. Many of my plants and propagation projects think it's summer year-round in my house! In this section, we will discuss how to create the ideal growing conditions for rooting cuttings and sprouting seeds indoors, because YOLO.

LIGHT

There is no such thing as a no-light houseplant, especially when it comes to propagation (unless it's plastic). All plants need light to grow and reproduce. Inside your plant's little cells are hard-working units called chloroplasts that are using energy from the sun (or your grow lights) to convert carbon dioxide and water into glucose (sugar) and oxygen. This is the process of photosynthesis, and it is what fuels new root and shoot growth for the plants you are propagating.

Just like most houseplants, your propagations will enjoy **bright, indirect light**. Many plant people have heard this phrase before. It may even be slightly triggering to some, as it's on nearly every tropical houseplant label. Yet, not many people understand it, leading many to unintentionally kill their plants due to a lack of proper light conditions. Raise your hand if this sounds familiar.

Let's gain a better understanding of what this phrase actually means. The *bright* part is simple: **put them as close to a bright window as possible**. For the fastest growth possible, there ideally should be nothing more than a pane of glass between your propagations and the outside world. Or, if you don't have a bright window, put them 1 to 2 feet (30 to 61 cm) beneath an LED grow light (depending on how bright your grow light is).

The second part of the phrase is *indirect*. This means to put them in a window where there the sun's rays do not shine directly down onto your plants for many hours at a time. One way to test if you are in the presence of direct sunlight is if objects blocking the path of the sunlight cast dark shadows. You want to aim for casting soft shadows with objects in indirect light!

Below is a list of the "best seats in the house" for your propagations (and any plants that enjoy bright, indirect light). Note that this list is for the northern hemisphere, and will be reversed if you live in the southern hemisphere. List not in any preferential order.

Propagation Locations:

North Facing Windows:
Medium to bright indirect light as long as they're completely unobstructed and plants are placed directly against window.

East Facing Windows:
Direct morning sun, then bright indirect light, then medium indirect light. Morning direct sun tends to be less intense than the afternoon, but if you find that your plants begin to burn, move them further away from the window, add sheer curtains, or move to another window.

West Facing Windows:
Morning indirect light, then bright indirect, then direct afternoon sun. Intense afternoon direct sun rays should be partially obstructed by sheer curtains, trees or buildings to prevent burning plants. If plants begin to burn, use a sheer curtain, move them further into the room or to a different window.

South Facing Windows:
Very bright, indirect light, with some direct sun possible depending on where you live. Right up against an unobstructed south-facing window may be too much light, so if this is the only type of window you have to work with, simply move your propagations further into the room or use a sheer curtain.

Under Grow Lights
Keep plants at the appropriate distance, depending on strength of the light and plant species (see next section).

Here is an example of what bright, indirect light looks like: My aroid collection consists of plants that grow in the dappled light of the jungle floor, in the wild. In my bedroom, they thrive under my north-facing skylight windows, where they never receive any direct sunrays. All plants have a direct view of the sky. Propagules also do well in this type of soft, medium to bright indirect light. I use my favorite pendant grow lights on the sides of my room to add additional light where the skylights end.

I use large pendant LED grow lights above my alocasia breeding area because the additional light helps them to flower. I have these lights here because they work well, and they're aesthetically pleasing. In my grow tent, I use grow light panels, which are not attractive, but effective and cover a large surface area.

GROW LIGHTS

I wouldn't be able to grow the large amount of tropical plants that I do, in Chicago year-round, without the use of grow lights. In fact, my entire Etsy shop and anthurium and alocasia breeding programs would cease to exist. I recommend full spectrum, bright white or warm white LED grow lights for growing and propagating houseplants. The purple (a.k.a. "blurple") LED lights also work well. Just note that your plants will appear purple all the time and your house may glow from the inside like a sexy late-night lounge.

I use everything from well-crafted hanging LED pendants with five-year warranties to inexpensive bulbs that screw into existing light fixtures to T8 shop-light-style bulbs that I hang from underneath wooden shelving. One of my favorite grow lights looks like a brightly lit robot starfish with movable arms on top of a tripod. Works like a dream but looks like a freak. Due to the houseplant boom and advancing technology, the options for modern, energy-efficient LED grow lights have exploded over the past few years, so you will not run out of choices. (See the Resources section at the end of the book for my grow light recommendations).

When it comes to operating your grow lights, buy an outlet timer and set your grow lights to run for 12 to 14 consecutive hours and be turned off the rest of the 24-hour day cycle. I used to use and recommend the manual outlet timers, but most recently I've fallen in love with the ease and convenience of smart plugs that connect to your phone and allow you to program your lights and turn them on and off from an app. I truly appreciate being able to do this now that my mother plants are so large. Before, every time I crawled through my jungle to reach the timers to reset or adjust them, I felt like I was lost in the Amazon.

The amount of light required for seed germination to take place depends on the plant species. Anthurium seeds do not typically need light to germinate (I've germinated them without), but germinating them under light does not harm them. Seedlings, however, need bright light and often require supplemental grow lights to grow thick stems and healthy roots and foliage while maintaining a compact size. As a bonus, sometimes the heat emitted from grow lights will help to warm the growing area, which can speed up growth.

The ideal distance to place your propagations and seeds from your grow lights depends on both the strength of the lights and the needs of your plants. Every grow light is unique in how much light it emits, which is why it's important to read the information provided by the light manufacturer. Many reputable brands provide graphics with suggested distances to place your plants, and companies even go so far as to provide you with a pamphlet when you receive the light that has an entire index of plants and suggested distances to place them from the light.

When I set up grow lights, I use knowledge and the powers of observation as my guide. I start by first learning about the plant I am propagating and the light I am working with and then I experiment and see what distances work for the plants (usually starting them at 1 to 3 feet [30 to 91 cm] away depending on the light I have . . . 4 feet [1.2 m] if it's an incredibly strong light). I check on the plants daily and adjust as needed.

However, I do own a **PAR meter**, a small light-monitoring device, which I recommend to anyone who would like to measure the exact amount of light emitted from their grow lights. A PAR meter measures **PPFD** or **Photosynthetic Photo Flux Density**. To put it as simply as I can, PPFD is the amount of photons (particles of light in the PAR spectrum) that your grow light is putting out per square meter (m^2) per second (S). Photons are measured in micromoles (μMol). Therefore, the units for PPFD is μMol/m^2/S. There is a lot of information from both the light manufacturers online and from experienced growers in social media groups around the world regarding the ideal PPFD for specific plants. You can do this! If you're not a math, science, or engineering person, you can always try the educated guess and observe route first.

HUMIDITY

Absolute humidity is the amount of water vapor in the air. Some plants need a lot of it for propagation to take place and some don't. Some types of propagation require high humidity and some don't. Every plant parent wants it, but only the best know how to get it. In this section, I'll let you in on all the secrets of the hobby's most highly sought-after environmental condition: high humidity.

Anyone who has attempted tropical houseplant propagation will tell you, the higher the humidity, the faster their cuttings root. I once spoke to a woman over social media who lived near a rainforest about her experience with propagation. She told me she could literally throw a philodendron cutting on the ground outside, come back a couple weeks later, and it would have rooted. For the rest of us not blessed with paradise-like conditions, we must depend on slightly less beautiful homemade propagation receptacles made of plastic bags and boxes to seal in humidity.

Measuring Humidity

As a houseplant owner, you can measure the humidity in your home, plant rooms, and inside your propagation boxes (discussed on page 68) with a little tool called a hygrometer. A hygrometer will give you a **relative humidity (RH)** reading. RH is a measure of the moisture in the air compared to the maximum amount of moisture the air can hold at a given temperature. The key thing you need to know is that **anything below 40 percent RH is considered low humidity, 40 to 60 percent RH is mid-range humidity, and above 60 percent indicates a high humidity environment for your plants.**

Propagation Humidity Requirements

Your plant's RH requirements during propagation depend on two factors: the type of plant you're propagating and the technique you're using. For example, a succulent native to Africa requires significantly less humidity for both growth and propagation than a plant native to the South American rainforest because the African succulent has adapted to life in the dry desert and can store water in its thick leaves for long periods of time without desiccation (drying out). RH of 40 to 50 percent is just right for propagating these dry-guys.

Dry desert plant = More water stored in leaves = Less humidity needed for propagation

On the contrary, humidity-hogs like the velvety anthuriums I grow, which are native to the rainforests of Central and South America, have thin leaves and need a humid environment to thrive. Otherwise, the leaves transpire (release water vapor into the dry air) too much too quickly and the plants droop and wither. Anthurium cuttings and other sensitive tropical plants thrive in temporary humidity levels of 80 to 100 percent during propagation. Humidity is even more crucial during propagation than throughout regular times of growth for a tropical plant because when you take a cutting, you are chopping it off from the primary way it obtains water—its roots! In the meantime, your plant's leaves will continue to transpire. Therefore, you need to consider the moisture levels in the air around your cutting prior to it growing

These little gadgets are hygrometers, otherwise known as your new best friends for propagation. They are inexpensive and you can buy them online. They measure temperature and relative humidity (RH). Every plant parent should own at least a couple.

An anthurium 'Silver Blush' I am propagating by stem cutting inside a propagation box with sphagnum moss in 90 percent RH.

An *Opuntia gosseliniana* pad I am propagating in cacti and succulent mix in 50 percent humidity.

roots. Hardier topicals such as pothos and hoyas can be propagated in water at any humidity level and do just fine, but they will still benefit from higher humidity levels to speed things up.

Tropical rainforest plant = Less water stored in leaves = More humidity needed for propagation

Propagation is stressful for cuttings. Therefore, humidity levels as close to 100 percent as possible are beneficial during propagation to ensure they get the moisture they require.

Extremely high humidity is always a requirement for tropical houseplant leaf cuttings that are lacking both roots and a stem to store water. Examples include begonia, African violet, and peperomia leaf cuttings. Give these leaf cuttings at least 85 to 90 percent RH at an ideal temperature range.

The chart below summarizes some of the most popular methods of houseplant propagation and their corresponding ideal humidity ranges that will encourage fast rooting:

PROPAGATION METHOD AND PLANT TYPE	IDEAL RELATIVE HUMIDITY FOR PROPAGATION
Leaf propagation of delicate topicals (i.e., rex begonias, African violets)	90–100%
Stem or petiole propagation of delicate tropicals (i.e., rex begonias, African violets)	75–100%
Nodes in a propagation box (i.e., philodendrons, *monstera obliqua*)	90–100%
Seed germination (i.e., anthuriums or alocasias)	80–100%
Air layering (i.e., fiddle leaf figs, philodendrons, anthuriums)	At least 50%
Succulent propagation (i.e., jade plant, haworthia, aloe vera)	40–60%
Simple stem propagation in water of hardier tropical plants (i.e., golden pothos, heartleaf philodendron)	At least 30%

Humidity Chambers

What do plastic food storage bags, dry cleaner bags, and plastic takeout containers all have in common?

They can all be upcycled to make humidity chambers. These are containers that you plant directly into, or place your pots inside of, to create humid microclimates for your propagules. The main requirements for your humidity chambers are that they're transparent or translucent so that light can get through to your plant, and the material should not be permeable so that warm, humid air gets trapped inside.

Additional ideas for humidity chambers include:

- Plastic zipper-top/food storage bags
- Making a mini-greenhouse above a container with a plastic wrap cover held up with toothpicks or skewers
- Clear plastic storage boxes
- Plastic takeout containers with clear lids like you see in the photo (the bigger the better)
- Large dry cleaner bags that you blow up like a balloon, then tie at the bottom or side around your plant
- A propagator box as seen in the photo above (a plastic base plus clear plastic lid often with vents to control humidity)
- Glass or plastic garden cloche
- Large glass jar placed upside down on top of your pot
- Grow tent or greenhouse cabinet (not technically "receptacles" but create microclimates for your plants and props)

[top] Plant babies growing inside a simple plastic box with air holes on top called a propagator (purchased online) and a DIY propagation box I made from a plastic takeout container. Both are helping keep the air around the cuttings more humid.

[bottom] A plastic zipper-top bag makes for an excellent humidity chamber because it's inexpensive, reusable, clear, and easy to crack open slightly if you need to allow some air to get inside in order to lower the humidity.

Prop Boxes

After you make your first **propagation box**, you become a certified plant person. Nothing says you've taken this hobby to the next level quite like a bunch of clear bins filled with plant cuttings and rehabs sitting under grow lights or stacked in front of your window.

A propagation box is simply a clear plastic bin of any size that one uses to maximize humidity while rooting tropical plant cuttings and rehabilitating tropical plants. The bin must be clear or translucent to allow light to enter. Depending on your humidity requirements, you can keep the box completely sealed (most common) or, partially cracked to allow for some airflow and to decrease humidity. You can even drill holes to increase air movement. Some people will burn holes in the sides or top of the plastic lid with a soldering iron. Personally, I keep my propagation boxes intact as I like to keep it as close to 100 percent humidity as I can inside, and crack the lid if necessary.

Anatomy of a Propagation Box

Clear plastic bin with lid

Space for stems and leaves to grow

Moistened rooting substrate layer (e.g., sphagnum moss)

Space for roots to grow

Drainage layer of perlite or LECA to capture excess water and nutrients

The Transition Out

Transitioning plants out of a propagation box is an art. You know how you feel when you step out of a hot tub into the cold air? I like to imagine that's how my plants feel when they are pulled out of the warm, moist propagation box.

When transitioning, some plants, such as hoyas with a waxy cuticle, are resilient to change and won't show any signs of shock or distress. Other tropicals, however, will wilt as a sign of protest. If you aren't careful in the way that you transition certain divas, such as alocasias, begonias, and anthuriums, they may potentially give up on life altogether. I really hate getting out of hot tubs too.

To ensure your most sensitive thin-leaved topicals are satisfied after their spa-like experience, try the following:

1. After cuttings grow roots that are roughly 1½ to 2 inches (4 to 5 cm) long, remove them from the prop box and repot them into a substrate of your choice.

2. If there is room, put them back—new pots and all—inside the prop box (or into a new humidity chamber) during this time of transition for another 1 to 2 weeks so that the plants do not have to adjust to the shock of a new substrate AND a change in humidity at the same time.

3. To keep the humidity as high as possible in the transition humidity chamber, keep a layer of moist substrate on the bottom under the pots. Or, simply mist the inside walls of the box with water.

4. After another week or two, transition the plants to a slightly lower humidity environment (ideally in the 70 to 80 percent RH range). Either keep them in that range permanently or, after another couple weeks, move them again to their permanent, even lower humidity area.

Pandora's box! This is what happens when you let your propagation box sit for months on end: an entire jungle grows. These plants all started off as nodes on bare stems or what plant collectors affectionately call **chonks.** The plants are all rooted in moist sphagnum moss, which is now covered in some green algae. Algae is perfectly normal and harmless at this level of growth.

[above] Total eclipse of the heart. This is what happened to my string of hearts propagules when placed on top of a faulty heat mat that ran too hot. This cup was placed on top of a plastic tray to protect it on top of the mat and they still cooked! Buying a heat mat with a thermostat can help prevent this from happening. When I repeated the propagation experiment without the heat mat, the hearts rooted like a charm. Because this has happened to me multiple times, I don't rely on heat mats much anymore. I may invest in more expensive ones in the future.

[page right, top left] Colored glass is ideal for water propagation.

[page right, top right] Cutting or soldering a hole into the bottom of a clear plastic cup is a great way to make a propagation vessel. These small dessert cups are all the rage for propagating small plants like hoyas.

[page right, bottom row] Make it fun. If it can hold water, you can propagate in it!

WARMTH

According to researchers at Michigan State University, the most ideal temperature range for your propagation media to encourage the fastest possible root and shoot growth is between 73 to 77°F (23 to 25°C). They recommend that air temperature be maintained between 68 and 73°F (20 to 23°C) when using bottom heat, such as when using a heating mat. However, if you don't have a heating pad at your disposal, air temperature should be increased to 77 to 80°F (25 to 27°C) to raise the temperature of your growing media. Air temperatures lower than this slow down shoot growth and promote root development. This being said, if you don't want to mess around with a heating mat, and you keep your home cooler than this (which is most likely the case for most people), your propagations will still grow roots, but be slower to grow new shoots than they would in a greenhouse environment. I live in Chicago. I know to be patient.

A seedling heat mat underneath your propagation box or seedling tray can be helpful to speed up the propagation process, especially during the winter months. Be extra careful if using a propagation box or humidity chamber with a seedling heat mat, however. If the substrate heats up too much due to a faulty heat pad or the heat pad being too close to your propagules or nodes, your nodes will start to look like burned asparagus. I know from experience.

PROPAGATION VESSELS

Picking a vessel to start your propagations in should be easy, accessible, and fun. It's a great opportunity to upcycle everything from cleaned out plastic takeout containers to clear plastic cups, glass sauce jars, and beer bottles. I'm not saying you should become a hoarder, but I am not saying you shouldn't become a hoarder, of propagation vessels. The more of these items you save, the more propagation vessels you're going to have on hand, while saving the environment. Just saying.

When it comes to propagating tropical plant cuttings in water, not all vessels are created equal. Colored glass reduces algae growth and keeps the water and vessel cleaner on the inside. For this reason, I collect brown glass vitamin jars for water propagation.

Cuttings with leaves that are narrower than the width of the propagation vessel opening may fall into the water. For this reason, some people make or buy shallow funnel-type propagation cones. These cones have a hole in the middle to insert cuttings, and rest on top of propagation vessels to keep your cuttings in place.

ROOTING AGENTS/ ROOTING HORMONE

Plants naturally produce different hormones that regulate different aspects of their growth and development. One class of hormones involved in cell division are called **auxins**. Indole–3–acetic acid (IAA) is the most common auxin in plants.

After you take a cutting of your plant, the auxins within your plant are responsible for facilitating the development of **adventitious roots**. Adventitious roots are new roots that grow during propagation. They include aerial roots, roots that grow from a node on a plant's stem, roots that grow from a leaf (such as a begonia leaf cutting), and those that grow on a corm or rhizome (such as in Alocasia corm propagation).

More auxins = More adventitious roots

It should come as no surprise then that if you want to speed up the rooting of your propagations, you should add more auxins. How can you get in on this extra auxin–action?

You may have to find a secret auxin dealer in a dark alleyway.

Or try searching the internet for "rooting hormone gel or powder." (Much less scandalous and exciting than the dark alleyway, I know.) Rooting hormone compounds, typically sold in gel or powder form, are made of synthetic (manmade) auxins. They additionally might contain a fungicide to prevent rot.

The rooting hormone gel I use is a water–based synthetic auxin gel made with a synthetic auxin, indole–3–butyric acid (IBA), along with mineral nutrients, trace elements, and vitamins for developing roots. To use rooting hormone gel, first pour some into a separate dish or container, to avoid contaminating the entire jar. (I'll be the first to admit I only remember to use a separate jar other than the cap of the container every other time). Then, dip the base of your cutting in, making sure to cover the areas where you would like new roots to grow. Next, place your propagule in the propagation medium. Finally, discard the used gel. It's as simple as that! I will warn you, however, that the gel typically has an intense odor.

The adventitious roots on my monstera mint NOID are growing like wild out of the nodes, thanks to auxins. Don't they look like sea anemones?

My favorite rooting hormone gel.

Rooting hormone powder.

Rooting hormone powders are very similar to the gels except that the auxin-based rooting hormones are mixed with talc. Just as with the gel, to use the powder, first pour a little into a separate dish or tray. Then dip your propagules into the powder and lightly tap to remove any excess. To help the powder to stick better, some people first dip their cutting into water prior to dipping it into the powder. However, studies show this leads to a more rapid deterioration of the rooting hormone.

As you'll see throughout the photos in this book, sometimes I use a powder and sometimes I use the gel, depending on what I have available. There are advantages and disadvantages to using either. Gel has an advantage in that it better sticks to the plant and is not rubbed off or washed off as easily as powder. Literature also shows that powdered forms of rooting hormones are generally less effective than liquid formulations applied at the same concentration. Auxin uptake by the cutting's base is often inhibited by the immediate removal of the powder from the cutting when it's put into the propagation medium. However, some people prefer rooting hormone in powder form because they're reportedly less toxic (they don't smell) and a bit easier to apply than a dark pink gooey gel. I recommend you experiment with both and see what you like!

NATURAL ALTERNATIVES

It is not uncommon to hear cinnamon recommended as a natural rooting agent. However, there is very limited scientific research to back up its effectiveness. The reason why some growers defend its use is that there is some anecdotal evidence that certain types of cinnamon help prevent fungal growth on plant cuttings, which could potentially prevent rot and aid in rooting.

Other natural alternatives include honey, seaweed extract, willow extract, aspirin, and aloe vera gel. However, while I always encourage experimentation, I believe it is more efficient to rely on scientifically proven methods, such as rooting hormone powder and ideal environmental conditions as discussed in this section in order to speed up plant propagation. I do really love cinnamon on my toast, though!

The one natural alternative to using synthetic rooting hormone powder that I do find most intriguing is putting a pothos cutting (or that of any other houseplant) in with your intended water propagation. The theory behind this is that the pothos cutting, which roots easily, may emit auxins into the water, and that will, in turn, help to speed things up for your intended plant cutting. Perhaps a pothos just makes your other cutting feel less lonely too. Propagation loves company.

OXYGEN

Oxygen helps promote root growth, as roots need oxygen to carry out respiration, a process that releases energy and allows them to grow and develop. Oxygen also allows for root nutrient uptake and prevents anerobic conditions, where bacteria and fungi grow like crazy and cause root rot. I know what you may be thinking: If you add more oxygen by adding fresh water or changing the water, would it help speed up propagation?

Yes, well, depending.

You see there is some nuance to this. Natural rooting hormones emitted by the plant circulate throughout the water during propagation and accumulate over time. Additionally, synthetic rooting hormone that you add in the form of gel or powder also helps to speed up propagation and circulates in your propagation water. Every time you change out the water to introduce more oxygen, you are flushing out all the accumulated rooting hormones. This is why *sometimes* plants will root *faster* when you don't change the water or change it less often.

However, there is a way to add oxygen and keep circulating rooting hormones levels high at the same time . . . while enabling you to be lazy (hells yeah)! Enter aquarium pumps and air stones.

Aquarium pumps that have air stones attached by tubing are an excellent solution to this problem as they introduce a constant supply of fresh oxygen to the water without you having to dump out the rooting hormones. Problem solved!

This being said, I only own one of these gizmos, and I am pretty laid back with my water propagations anyway, so most of them get their water changed and oxygen refreshed whenever I remember. I like to tell myself I'm purposefully letting the rooting hormones accumulate.

Having the right tools and materials on hand for propagating your houseplants makes the task so much easier. Just think of how much better your props are going to grow for you after the plant supply shopping trip you're about to go on (but I refuse to take the blame for). Now that you know all the elements needed for success, let's get to the fun part: the propagation methods, step by step (ooh baby)!

I'm using an inexpensive aquarium air pump with a silicone tube connected to an air stone to add a constant supply of new oxygen to the water in my propagation vessel.

Chapter 4

PROPAGATION METHODS

Everyone remembers their first time.

The first time I learned about houseplant propagation many moons ago was when a friend posted before and after pics on social media. In the before pic, she presented a plate filled with succulent leaves. She explained that she plucked each one of them off a single plant with the intention of turning each into a new plant.

The after photo showed the same leaves, but now they each had tiny succulent babies attached. SHE DID IT. Was she a witch?

I was enthralled. I wanted in.

Could it really be that easy? Just lay the leaves out on a plate and wait? I doubted it. I didn't understand succulents.

I hastily bought a succulent of my own, pulled it apart, and proceeded to do what any plant mom with good intentions would do—provide the leaves with lots of water, weekly, so they would grow. I even added fertilizer for an extra boost. I kept them on my living room shelf where they received only a few hours of afternoon light peeking in from my city balcony each day.

Let's just say I wasn't quite ready for the main stage yet. Those leaves and my confidence in my own ability to propagate plants died a mushy, miserable death. However, although succulents and I have a complicated relationship to this day, I have redeemed myself with many a successful leaf propagation since then. (This would be an awkward introduction to a how-to section if that weren't the case.) I learned the hard way that sometimes less is more when it comes to succulents, and more importantly, it pays to do a little reading prior to imitating something you see on social media.

And that's where this chapter comes in. I am going to walk you through, step by step, how to make your own free plants in the easiest and most efficient ways possible so you can avoid repeating my great succulent leaf tragedy. My guarantee is that after reading this part of the book, you will be just as enthralled as I was upon seeing my friend's succulent leaf post for the first time.

Let's start with the tropical houseplant propagation techniques first for all of you philophiles, ficus fanatics, and hoya heads. Then, we'll move on to cacti and succulent methods for the plant parents with desert dreams. The final part of this section will focus on trending techniques that are specific to particular plant species.

TROPICAL HOUSEPLANT PROPAGATION

What did the propagule ask its estranged mother plant?
"Why did you cut me off?"

All right friends, it's time for STEM class.

Literally.

Stems are not the sexiest part of a plant, and so it's easy to overlook their importance, but this section is their time to shine. Stems not only transport water and nutrients, provide structure, and hold leaves, but they contain special tissue that gives them new root- and shoot-generating superpowers! The stems of some plants also contain nodes where adventitious roots readily grow from, which is why stems are voted the "most likely plant part to be propagated" year after year. Finally, some stems are hairy. So, if you're into that kind of thing, they may be right up your alley.

Stem propagation has a high success rate and is relatively straightforward, making it a low-risk, high-reward activity. If you need a little self-esteem boost or pick-me-up, go on and propagate your pothos by stem cutting. In my opinion, nothing says "I'm a winner" like the successful rooting of a houseplant.

While your stomach may flip the moment before you take that initial snip into your plant baby's stem, there's something addicting about it once you complete the initial cut. If you are going to put the time and effort into making one baby from a mother plant, you might as well make another, or five others! After all, if the mother plant has a healthy root system, she will grow back just fine. And if you're lucky, she will grow back extra lush, as cutting the stem often awakens dormant buds on the plant, just below the cut, causing them to branch out.

Prior to beginning stem propagation, you have two major choices to make:

1. What substrate are you going to root your stem cuttings in?

2. What propagation vessel are you going to use?

If you need extra help with these decisions, simply flip back to the last chapter.
Next, choose your own propagation adventure:

1. If you would like to learn how to make baby plants in the most simplistic way possible, using a glass of water, continue reading below.

2. If you want to learn how to use a propagation box for your stem cuttings, skip to page 84.

3. If you chose to give up now, and get another hobby, turn to page 190.

I made a second fiddle leaf fig plant for myself by simply cutting the top off my existing one and sticking the top cut in moist sphagnum moss for a few months. I never let the moss dry out. Figgin' easy!

STEM PROPAGATION IN WATER

Featuring Jade Pothos (*Epipremnum aureum*)

Taking a plant stem cutting and sticking it in a vessel of water until it grows roots is the OG method of houseplant propagation. It's easy, affordable, and intuitive. The best part of stem propagation in water is that it works with stems of all sizes and textures. Whether you want to make more plant babies from one of your trailing plants with itty bitty stems, a monstera with a thick chunky-monkey stem, or a fiddle leaf fig or rubber tree with a tough, woody stem, water propagation is going to be a great option.

Here we review how to use water to make more babies of the pothos plant, because we are all less than six degrees of separation from a pothos. At one point or another, everyone either owns, knows someone who owns, or works at an office that has this classic houseplant.

Supplies Needed

- Sterilized cutting tool
- Rooting hormone gel or powder (optional)
- Clear vessel

Timeline

- 3–6 weeks for roots to develop on a pothos in water
- 2–8 weeks for other houseplants, depending on the plant species and environmental conditions

Ideal for

- Most vining and trailing houseplants, such as species and varieties of pothos, philodendron, monstera, syngonium, hoyas, lipstick plant (*Aeschynanthus*), tradescantia
- Ficus species such as fiddle leaf fig (*Ficus lyrata*) and rubber plant (*Ficus elastica*)
- *Schefflera* spp. (umbrella plant)
- *Aglaonema* spp. (Chinese evergreen)
- *Pilea peperomioides* (Chinese money plant)

nodes

internode

1

2

Instructions

Step 1: Prepare Your Vessel

Find a transparent vessel to put your cuttings in. I love to use cute small bud vases that have tapered necks like the ones you see here, for small cuttings. You can buy them online or at a craft store. Most recently I found a brand of rice pudding that comes in the perfect-sized propagation jars at the grocery store, so now I must make the ultimate sacrifice and eat large amounts of rice pudding . . . for my plants. Some people drink beer for the amber glass bottles. We all must make sacrifices. Clean out your jar with soap and water or rubbing alcohol to sanitize it before using.

Fill the vessel up with fresh water ¾ of the way. Tap water should be just fine in most cases, but some people prefer to use distilled or filtered water.

Step 2: Take Your Cutting(s)

Decide how long you'd like your new cutting to be, then cut off a piece of the stem at least an inch below any node, along the internode, as shown. Cut so that there will be enough space on the cutting for roots to grow. As a reminder, nodes are the areas along the stem that contain dormant axillary buds. Don't be scared. Most plants love to be pruned and will grow back if they have a solid root system, water, and light. Take as many cuttings as you'd like, but remember to leave at least one node on the mother plant to regrow.

3

well in moderate light; every species has a preferred range. Humidity and warmth also help them to grow faster.

Pour at least some of the old water out and add fresh new water weekly (or when you remember). New water will supply fresh oxygen to the plant's tissues.

Step 5: Pot Them Up

Once the roots are at least 1½ inches long (4 cm), your once-baby cuttings are ready to be placed in a grown-up plant pot. Choose a pot that is no bigger than 1 to 2 inches (3 to 5 cm) larger in diameter than the new roots. To minimize root rot, use a pot that has a drainage hole.

Next, pour an initial layer of 2 to 3 inches (5 to 8 cm) of well-aerated tropical plant potting mix (I like to mix together 20 percent perlite plus 80 percent tropical plant potting mix straight from the bag for a pothos) or aroid mix into the pot. Place your cutting into the pot with the roots resting on top of the initial layer. Then, fill in the area around the roots with more potting mix. Continue to fill the pot until all the roots are completely covered. Leave at least ½ inch (1 cm) of space between the top of your potting mix and the rim of the pot for water to collect (and not spill over) when you water it. Water the rooted cutting thoroughly until water drains out of the bottom of the pot.

Tip

Place a small piece of window screen or houseplant pot screen over the drainage hole to prevent the potting media from spilling out of the drainage hole.

Step 3: Plop in Water

Applying rooting hormone powder or gel to the cut end of the cutting prior to putting it in water is not necessary, but it will help speed up the rooting process. Dip the entire base in, from where you made the cut up to, including the lowest node. While the water will wash some of it off, much of the hormone will still cling on, and that which doesn't will still circulate in the water around the plant (until you change the water).

Arrange your cuttings in a little bouquet in your vessel. Ensure that the bottom-most nodes are submerged under the water; from there, roots will grow. Keep all leaves above the water surface to prevent the leaves from rotting. It is perfectly okay—and even encouraged—to pick off any lower leaves that are submerged in the water. Other ideas to help keep the leaves above the water include using a propagation cone, plastic wrap with a hole, tape, or rubber bands across the top of your vessel.

Step 4: Set the Mood

What, you didn't think you had to set the mood for asexual propagation? If you want your plants to reproduce, you must set them up for success. Give your cuttings bright, indirect light (see page 62). Many cuttings will also root

Step 6: Aftercare

Place your new pothos in a warm spot that receives bright, indirect light, and water it whenever the soil is nearly dry, or in other words, when the top few inches (about 5 cm) of soil are dry. I recommend fertilizing tropical houseplants year-round if they are actively growing.

This 'El Choco' needs el choppo!

NODE PROPAGATION IN A PROPAGATION BOX

Featuring *Philodendron rubrijuvenile* 'El Choco Red'

Why did one philodendron network with another to enable a successful propagation?

Because it's all about who you node.

In chapter 3, we reviewed how to set up a propagation box and how to use one to seal humid air around your propagules. With the use of ye ole' prop box, you can raise the relative humidity surrounding your cuttings to near 100 percent, creating the perfect jungle-like microclimate for rapid root and shoot growth. The beauty of a prop box is that if you set it up correctly, keep it continuously moist inside, and give it enough light and comfortable room-temperature levels of warmth from the outside, any healthy propagules you put inside will likely take root. Whether you're trying to root corms, rhizomes, stem cuttings, or nodes (as we will discuss here), it doesn't really matter. What goes in the prop box, roots in the prop box.

Climbing philodendrons, such as *Philodendron erubescens* 'Pink Princess', *P. verrucosum*, and *P. rubrijuvenile* 'El Choco Red', make for stunning houseplants, especially if you are growing them inside the humid environment of a greenhouse cabinet or grow tent, or you're using a humidifier to keep these aroids happy. When provided with proper nutrients, humidity, and light, they can grow quickly and their leaves can become humongous.

However, life is not perfect, and neither are our plants or our growing conditions. As a result, sometimes these plants get tall, but not lush. These climbers are notorious for their bottom leaves falling off, leaving behind a bare bottom stem and only leaves on top, making the plant resemble a lanky tree tied to a moss pole or plank. Or, maybe your plant is recovering from a pest outbreak and it just looks outright hideous. Whatever your motivation, let's discuss what to do when you want to chop your beloved climbing aroid to bits and start fresh and new (with many more plants).

Ideal For

- Climbing philodendron species (e.g., *P. erubescens* 'Pink Princess', *P. rubrijuvenile* 'El Choco Red', *P. sodiroi*) and similar plants
- *Monstera deliciosa* and other similar monstera species
- *Monstera obliqua* (they produce long runners you can propagate this way)
- *Amydrium medium* (also produce runners you can propagate this way)

Supplies Needed

- Sterilized cutting instrument
- Clear plastic box with lid
- Sphagnum moss
- Perlite or LECA (optional)
- Rooting hormone powder/gel (optional)

Instructions

Step 1: Set Up Your Propagation Box

Prepare your propagation box using the substrates of your choice. My personal go-to is placing 3 to 4 inches (8 to 10 cm) of moist sphagnum moss on top of a couple inches of chunky perlite or LECA to act like a water-holding reservoir on the bottom. If you don't have any one of these substrates, it's perfectly okay to use *only* sphagnum moss or *only* perlite or LECA. Water so that not only is all the substrate in the box moist (sphagnum moss should ideally always be as moist as a wrung-out sponge), but add extra water so that there is about ½ inch (13 mm) of water on the bottom of the box to act as a reservoir.

Step 2: Isolate the Nodes

Cut your plant's main stem into pieces at the middle of the internodes (the spaces between each node). Each piece should contain a single node at its center point. If you are working with a leafless runner, as is often the case with a *Monstera obliqua*, you will still make your cuts in the middle of the internodes. Never cut on top of—or too close to—a node or you may risk cutting into a dormant bud. You will create little sticks, which plant collectors call chonks or wet sticks.

Step 3: Rooting Hormone Compound

Dip the nodes in rooting hormone powder or gel. Make sure to cover the brown bumps, which are aerial roots beginning to form. These roots will extend and grow into terrestrial roots inside the propagation box.

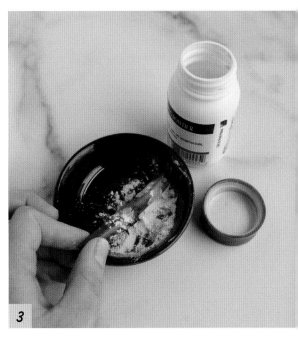

4A **4B**

Step 4: Put the Node Cuttings in the Box

Rest the node(s) right on top of the substrate inside of your propagation box and close the lid. You can partially bury the nodes or simply leave them laying on top.

Now, whisper:

Goodnight moss. Goodnight nodes. Good night props in your prop box abode.

Step 5: Transplant Your Baby!

Once the roots are at least 1½ inches (4 cm) long, carefully remove the sphagnum moss from the roots and pot up your philodendron in a chunky aroid mix. Refer back to the substrate section of chapter 3 for more information on how to make an aroid mix. Water the pot thoroughly and put it in bright, indirect light in a warm location.

Step 6: Philodendron Prop Aftercare

For the first week or two after potting up your new baby, keep it in a high humidity environment (70 to 100 percent RH). The high humidity helps the baby adjust to a new substrate without stress, and it will grow more roots, faster. This is not a necessary step by any means, but it helps. After this short adjustment period, *Philodendron* 'El Choco Red' will thrive in an environment that remains humid (50 to 60 percent and above at all times), and it enjoys having a moist moss pole or wooden plank to climb up. I grew the specimen you see in the photos to a large size from a small plant by placing it a few feet away from a large LED pendant grow light in 60 percent humidity. It climbed up a moss pole I would mist a few times a week.

Other Options

While in this section I illustrate how to root cut pieces of philodendron stems with a node on them (a.k.a. nodes/chonks/wet sticks) in the propagation box, know that you could *just as easily root trailing plant stem cuttings with leaves still on them.* For example, simply take a cutting of your trailing heartleaf philodendron or pothos stem, remove the bottom leaves to expose some nodes, and insert it in the moss inside the box to root. If doing this with a hoya cutting, there is no need to remove the bottom leaves; the roots will emerge along the stem. Easy peasy propageesy!

5

Petiole

PETIOLE PROPAGATION

Watermelon Peperomia (*Peperomia argyreia*)

Some plants, such as peperomias, certain species of begonias, and African violets, can sprout entirely new baby plants directly from the bottom of their petioles after they've been cut from the mother plant. It's truly a peculiar thing to witness, and the babies look like they are growing from the wrong spot. The last place you'd ever expect a bunch of fully intact baby plants and roots to sprout from would be the end of a dainty petiole. And it's for this reason that every plant parent must give this method of propagation a try. It's also incredibly easy.

As shown above, the petiole is the stemlike stalk that attaches the leaf to the stem. On aroids, such as philodendrons and monsteras, the petiole is thick and strong to support the giant, thick leaves. Propagation is not possible from these petioles because the cells there are not totipotent, which is why we use stem propagation with these plants. It's the plant genera and species with totipotent cells in their petioles that are more likely to reproduce using petiole cuttings.

Petiole propagation is very straightforward. The tricky part is keeping the mother plant alive long enough to be able to propagate it (especially if you're dealing with a fancy begonia or a delicate wrinkle-leaved *Peperomia caperata*). If you know, you know. In this section, I will use the popular watermelon peperomia as an example, which is a bit hardier than its super sensitive and wrinkly *P. caperata* counterpart.

What substrate you choose to propagate the petiole cutting in is up to you. Water, as always, is the simplest method. However, be aware once again that the roots that develop will be acclimated to water so when you transfer the babies to potting mix you may need to give them a little extra TLC in the form of added humidity and well-aerated soil while they adjust to that soil life. Perlite, vermiculite, and Fluval Stratum or a mix of these are also excellent options for petiole propagation because they provide physical structure for the petiole to stand upright in, space between the particles for oxygen, and are easy to remoisten.

Preparation

Begin propagating a few days after a thorough watering so that the leaves and petioles are fully hydrated.

Timeline

- Approximately 3–4 weeks for roots to develop
- 2–3 months until there are baby plantlets at the base and the babies can be potted up

Ideal For

- Peperomia species
- *Begonia rex* and other begonia species
- African violets (*Streptocarpus* spp. Sect. *Saintpaulia*)
- *Sinningia speciosa* (gloxinia)
- *Episcia* species

Supplies Needed

- Sterilized cutting shears
- Rooting hormone compound

If propagating in water:
- A propagation vessel for water propagation (Thin test-tube propagation stations work great for this method because the leaves can't fall in.)

If using a semi-hydroponic setup (as shown):
- Small plastic container or cup with drainage holes
- Wider cup or container without drainage holes to act as a water reservoir
- Perlite
- Humidity enclosure

Instructions

Step 1: Take Your Petiole Cuttings

Choose a firm, healthy leaf and petiole to cut with a sharp, sterilized cutting instrument. If you're working with a plant that has long petioles, cut your petioles so that the cutting is 2 to 3 inches (5 to 8 cm) long. If your plant has short petioles, cut as close as you can to the main stem to get as much petiole as possible.

1

Tip

If working with a *Pilea peperomioides* (Chinese money plant), you must get a little sliver of stem at the base of your petiole to propagate it. Because a piece of stem is included in the cutting, it is technically a stem cutting but it is rooted in the same manner as a petiole cutting. Retaining a piece of stem tissue on the bottom is not necessary for other peperomia species, begonias, or African violets.

2

Step 2: Plant Your Cuttings

Dipping the petiole in rooting hormone compound prior to planting is optional.

For simple and easy water propagation, stick the petiole cutting in a vessel of water. However, I highly recommend propagating in perlite in a semi-hydroponic setup with a water reservoir as shown in the photos. It may sound a little complicated, but once you understand what's going on, it's straightforward:

1. First, poke holes into the bottom of a plastic takeout container or dessert cup, or use a nursery pot with holes on the bottom.
2. Fill the container with perlite.
3. Plant the petiole cuttings (I also planted some leaf cuttings; instructions on how to do that in the next section) into the perlite.
4. Set the entire container into another, wider container without a drainage hole filled with enough water so the planted container is submerged to just below where you think the new roots will grow on your propagations. Fill the bottom, wider container with 1 inch (3 cm) of water.
5. Water or mist the top to moisten the perlite on top as well.
6. The perlite will slowly wick up the water from the reservoir below via capillary action throughout the propagation period. When the water in the bottom container (the reservoir) dries up, add more.

This is an extremely effective and sterile way to propagate plants. It also works very well for anthurium stem propagations, which are notorious for rotting.

Tip

Trimming the top half of the leaf off on your petiole cutting may help the cutting focus its energy on producing roots, faster.

3

Step 3: Maintain High Humidity

If propagating in media other than water, place your propagation vessel in a propagation box, propagator, plastic bag, or similar enclosure so the substrate won't dry out too quickly and the humidity around the cuttings is kept high.

Petiole Propagations

4

⌒ Semi-hydroponic perlite propagation results.

⌒ Water propagation results.

Step 4: Pot It Up!

Your prop is ready to be potted into a well-draining potting mix when the roots are at least 1 inch (3 cm) long. If you wait a bit longer, that's okay too. These pictures are the results after 3 months. Keep the original mother leaf and petiole attached if it still looks healthy. Once it starts to die, feel free to cut it off at the base of its petiole. The babies will have their own roots!

When potting up peperomia propagations, making your own well-draining potting mix consisting of 70 to 80 percent tropical plant potting soil straight out of the bag mixed with 20 to 30 percent perlite is recommended. Just eyeball it. The goal of adding perlite is to add additional air pockets to the soil. Give the pot a thorough watering.

Step 5: Aftercare

Provide your peperomia with bright, indirect light. It will appreciate humidity around 50 to 80 percent if possible but will tolerate lower humidity. Water when the plant has dried out halfway.

LEAF VEIN PROPAGATION

Featuring *Begonia rex*

You're so vein. You probably thought this section was about you . . .

But it's about leaf vein propagation, sorry. And if you thought petiole propagation was wild, things are about to get much crazier up in here. Those same dainty plants we spoke about in the last section—begonias, peperomias, and African violets—have an amazing superpower. If you slice one of their leaves open, they can grow new baby plants from along the cut.

This method of propagation is not only incredible to witness, but when used along with petiole propagation, it enables you to utilize most of the mother plant's tissues for propagation. The result? More babies! The downside to leaf vein propagation is that leaves are delicate. Not all leaves will root and sprout, and some will completely rot. If you go into it with realistic expectations—and a propagation box—you're all set.

Timeline

- 2–3 weeks for roots to develop
- 6–8 weeks for plantlets

Ideal For

- Rex begonia (*Begonia rex*)
- Iron Cross begonia (*Begonia masoniana*)
- Peperomia species

Supplies Needed

- Humidity enclosure
- Sterilized knife, X-acto knife, or razor blade
- Cutting board
- A shallow container at least 2 inches (5 cm) deep (plastic takeout containers work well)
- Propagation mix (peat moss + perlite) or substrate of your choice
- T-pins or floral pins, helpful but not necessary if using the exact method illustrated here
- Tropical plant potting mix
- New pots for your babies

Preparation

Water your plant a few days prior to taking your leaf cutting so that you begin with a well-hydrated leaf. This is particularly crucial for tropical plant leaves that tend to be thin and have limited water storing capacity, making it quite difficult for them to survive without roots.

Instructions

Step 1: Remove the Leaf

Choose a mature and healthy leaf.

Step 2: Slice It Up!

Slice the leaf into at least two wedges, but I recommend at least four. Go hard or go home. The goal is to slice across major veins in the leaf. From those veins, begonia babies will grow.

Step 3: Plant

First, fill the container you plan on planting your leaf wedges in with moist substrate. I use a mix of fully moistened peat moss and perlite. Alternatively, fine moist perlite alone, or vermiculite works very well for this application because of their fine texture, ability to mold around a leaf, and porosity. Make slits in your substrate with a chopstick, seed dibber, skewer, or similar thin, pointy object in the exact locations where you'd like to plant the leaves.

Next, dip the cut, bottom edges of the wedges in rooting hormone (optional) and plant them approximately ½ to 1 inch (1 to 3 cm) down into the substrate inside the slits you made. With your tool or fingers, fill in the gaps along the sides of each leaf with substrate so the leaves are secure and standing upright. If you have any remaining leaf tissue after you plant your cut wedges, pin it down in your substrate as well. Every part of the leaf that is cut and contains exposed veins is fair game here.

No tissue left behind!

Tip

Another way to propagate a *Begonia rex* from leaf veins is to keep the leaf intact but make little ½- to 1-inch (1 to 3 cm) slits along the veins on the underside of the leaf, right below where the veins split. Then, lay the leaf, underside down, onto the substrate and use little T-pins or floral pins to hold it down (ideally), or tamp down the edges of the leaf simply by placing substrate or rocks on top of them. New plantlets will form from the slits.

Step 4: Humidity, Humidity, Humidity!

These thin, sensitive leaf cuttings will not survive very long without being in a high humidity environment because they will transpire and lose moisture to the air. Seal the humidity in by enclosing your container in a humidity dome as discussed in chapter 3. Suggestions include a clear plastic box, plastic zipper-top bag, or make your own mini greenhouse by placing plastic wrap over the entire project, with the roof held up and away from the leaves using skewers.

Step 5: Observe the Growth

After 6 to 8 weeks, new plantlets typically appear. Once they are around 3 inches (8 cm) tall, you are safe to plant them in their own pots.

5

Step 6: Pot Up the Babes

Pot your new baby begonias in a well-draining but water-retaining tropical plant potting mix. You can separate the new plantlets by gently pulling them apart. Make sure the pot is cute, or the babies won't be satisfied (kidding). Use a small pot if planting individual props. The pot you choose should only be 1 to 2 inches (3 to 5 cm) wider in diameter than the root mass.

Step 7: Aftercare

As with all newly potted up propagations, your begonias will enjoy a little extra humidity while they get acclimated to their new substrate. Begonias like bright, indirect light (what else is new?). Water when the top couple inches of soil start to dry out, or in other words, when they're partially dry. After they're acclimated to their new pots, relative humidity levels above 50 percent are how they will thrive, but they will put up with lower humidity levels in your home if they have to. Just don't ask them to fill out a satisfaction survey.

6

DIVISION

Featuring *Goeppertia roseopicta* 'Rosy' (*Calathea*roseopicta* 'Rosy')

*This plant was reclassified as a goeppertia, along with many other—but not all—calatheas.

If you are getting nervous about dividing your houseplant into multiple houseplants, you likely are overthinking it. I know from experience. It's actually much easier than you think! If you're an outdoor gardener, you already know this.

When I first started out with houseplants, a friend of mine told me that every spring he divides up his calathea into multiple plants to give away to friends. I listened with amazement and fear, and wondered how he did that without damaging the plant or its roots.

I had no idea just how resilient plants were. Goeppertias (or calatheas as many of you know them by) die for 1 million different reasons. They are total divas and require consistent humidity or they turn crispy. Some people even go so far as to water them with rainwater, reverse osmosis water, or water them and then let them sit out overnight, to reduce the amount of chlorine and fluoride the plant receives. Also, they will get spider mites and die if you look at them funny.

But because you divided them is not typically one of those 1 million reasons.

In this section we will review how to divide your geoppertia, calathea, or any tropical plant that has multiple plants growing in one pot connected by thick, underground stems, called rhizomes.

It's important to note first, however, that any houseplant that is a combination of multiple houseplants in one pot can be easily separated into individual plants without having to break apart any underground stem tissue. Although these lush, bushy plants appear as though the vines are all connected at the root ball, it's an illusion. When you see bushy pots of hoyas, pothos, string of hearts, and trailing philodendrons, they are likely composed of numerous individual plants combined into one pot in order to sell. Dividing these plants is simply a matter of pulling them apart. In this section, however, we are going to discuss how to divide plants that you need to physically break or cut apart to divide because each clump is connected to the other, underground.

Timeline

- 15–30 minutes

Ideal For

- All tropical houseplants that spread underground via rhizomes, including, but not limited to, many plants in the *Marantaceae* family such as calatheas, goeppertias, stromanthes, and ctenanthes
- Peace lilies (*Spathiphyllum*)

Supplies Needed

- Sterilized cutting instrument
- New pots for divisions
- Fresh tropical plant potting mix (I recommend tropical plant soil plus extra perlite)

Instructions

Step 1: Remove Plant from Pot

Step 2: Remove Some of the Old Soil and Loosen the Roots

Use your fingers to remove enough soil to gain visibility over what is likely a hot mess root entanglement situation.

Step 3: Cut or Break Your Plant into Sections

First look for natural clumps within your plant. Your plant will reveal to you where you should make the cut or break it apart. You will notice that your bushy plant is naturally grouped into sections, which look like little bunches, connected by a thick underground rhizome.

While you inspect the roots, if you notice any areas of brown, mushy rotted roots, remove them.

Snap or cut each section you want to divide away from the mother plant. Look for sections that have full, mature root systems. They will do the best on their own as divisions. If the rhizome is extra thick, or you want to be clean and cautious, use a sterilized scissor or knife to cut the rhizome. Try not to break a lot of roots as you gently pull the plants apart. However, if you do break a few roots, no one will call the plant police and your plant will recover.

Step 4: Pot Up Individual Plants

Give each plant its own pot with well-draining tropical plant mix. For calatheas and goeppertias, I use tropical plant potting soil right out of the bag with about 20 to 30 percent extra perlite mixed in. Water thoroughly.

Step 5: Aftercare

Your divided plants may or may not look sad and droopy for a few weeks or even a couple months after propagation as they acclimate to being off on their own in this big world. If this happens, give them time, bright indirect light, warmth, and lots of humidity. Cut off any leaves that completely crisp up. Only water your calathea/goeppertia when it's almost dry. It should soon recover, and you'll have a new plant!

Pilea peperomioides is in the genus *Pilea*, and its species name, *peperomioides*, means "peperomia-like." This does not mean pepperoni-like, although the leaves are round like pepperoni. If you've ever seen what a plant from the genus *Peperomia* looks like, this makes complete sense!

OFFSET/PUP PROPAGATION

Featuring *Pilea peperomioides* (Chinese money plant)

Some houseplants literally beg to be propagated, such as the spider plant and *Pilea peperomioides*.

How?

By making irresistible pups for you!

Both these plants do all the hard work for you and create little baby clones that remain attached to the mother plant, called pups or offsets. Even the strongest attempt to resist the temptation to remove a cute little pup from the mother plant usually fails in the end.

Pilea peperomioides is a wacky-looking plant with many nicknames: the UFO plant, the Chinese money plant (the leaves are round like coins), and, my favorite, the friendship plant. I'm not sure if anyone truly reads plant book introductions, but if you read the one to this book, I mention how propagation can strengthen interpersonal bonds. Through the gracious act of passing along a free plant, you can strengthen relationships and meet new people. Well, if there was any plant on the market that would aid in this quest the most, this is it! Not only does *Pilea peperomioides* produce baby after baby, but not many people can resist its adorable shape and easy-going nature.

Okay, I lied. If there was any plant propagation that would help you win the hearts of others, it would be gifting a cutting of the extremely rare variegated monstera 'Devil Monster' that just sold in Asia for $38,000. Please note, I am open to making friends with you if you plan on doing this.

But I digress . . .

Timeline

- 15–20 minutes for division
- Approximately 2–4 weeks for the pups to establish themselves in new pots

Ideal For

- *Pilea peperomioides* (Chinese money plant)
- Spider plant (*Chlorophytum comosum*)
- Bromeliads

Supplies Needed

- Sterilized cutting shears
- 2–3.5-inch (5–9 cm) pots for each pup
- Rooting hormone (optional)
- Well-draining potting mix

1

Preparation

It's ideal to wait until the pups are at least 2 inches (5 cm) tall prior to separating them, although with extra care, they may do just fine if you propagate them at a slightly smaller size.

Instructions

Step 1: Remove Mom from Her Pot

Slide your plant out of its pot.

Tip

If your *Pilea peperomiodes* grew so many babies that it is now root bound and stuck, run a butter knife around the inner edge of the pot to loosen the outer edge of the root ball away from the pot. Then, holding the pot upside down and holding the plant, tap on the bottom of the pot with the heel of your palm while pulling the plant out. If your plant will still not come out using this method, you may have to smash your pot with a hammer to get mama out!

2

Step 2: Separate the Babies

Separate each of the pups either by breaking them off with your hands at their base (the quick and easy way) or, you can be more precise and cut them off with a sharp knife at their base. Ideally, all the pups have some roots already. If they don't, you have two choices. One, you can treat the rootless pups like the pups that do have roots, plant them directly into potting mix, and hope for the best (honestly, this is what I do and it usually works out just fine). Or, you can play it safe and put the rootless babies in a glass of water for a couple weeks until they have some roots. After you see roots (even tiny ones less than 1 inch [3 cm] long), you can pot them. These babies are resilient!

Step 3: Pot Up the Pups

Whether or not they have their own roots, pot each one in its own pot with a well-draining potting mix. A tiny 2 to 3.5-inch (5 to 9 cm) pot with a drainage hole works great to start. *Pilea peperomioides* are so hardy that the pups will survive and grow roots of their own without any special treatment prior to planting.

Step 4: Water Thoroughly

Water the potting mix thoroughly.

Step 5: Aftercare

Place the babies in very bright, indirect light, and treat them as you would the mom. These are tough little kids; they can handle it! Water them when the soil mix is almost dry. In 2 to 4 weeks, they will be fully acclimated and growing longer roots.

Tip

When you pot a plant for the first time, to ensure the potting mix gets fully saturated, it's a great idea to bottom water. Sit the entire pot in a saucer or container of at least 1 inch (3 cm) of water and allow the soil mix to soak up the water from the bottom through the pot's drainage hole. By the end of the day, the entire pot of soil mix will be fully saturated.

3

Inside this plastic covered ball of moist sphagnum moss, there are philodendron roots starting to take off.

AROID AIR LAYERING

Featuring *Philodendron erubescens* 'Pink Princess'

Air layering is the perfect propagation technique for the risk averse. It is the "try it before you buy it" of propagation methods. Or, shall we say, "prove it before you remove it." During air laying, you *first* encourage your mother plant to grow roots from a node and/or stem tissue, and *then* you take the cutting. Air layering can be performed on woody-stemmed ficus species, including the popular fiddle leaf figs and rubber plants, but the technique looks slightly different. In this section, we will discuss how to air layer aroids, such as philodendrons and monsteras. This propagation technique is one of my favorites for making more of some of my most cherished plants.

I bought a *Philodendron* 'Pink Princess' a couple of years ago as a tiny plant with a single goal in mind: to grow it as big as botanically possible. I potted it with a moss pole that I kept moist at all times. I grew it under a bright grow light in a grow tent at 75 percent relative humidity. I did all the right things—perhaps too many of the right things. Because quickly, I ran into a major obstacle: the plant hit the top of my grow tent.

I had to decide whether I was going to move my family inside a giant glass conservatory, cut hole in my grow tent for the plant to peak out of, or, more practically, cut and propagate the top and start it as a new plant. From the new plant I could then continue my quest for 'Pink Princess' world domination. In the plant collecting world, after chopping the top off a plant like this to propagate it, we refer to the top as the top cut and the bottom as the butt cut. (I am thoroughly disappointed that I cannot take credit for making up the term butt cut.)

After my husband rejected my plea to move into a conservatory for more vertical space, I needed to decide how I wanted to propagate my 'Pink Princess'. Should I just chop her top off and throw it in a huge vase of water or moss and then try to find a place to put it for a month while it grew roots without my three kids knocking it over? Or, could there be a more efficient and space-saving way to propagate her?

Enter air layering.

There are a few key reasons why I love this method of propagation and they all stem from the fact that you root the cutting before you cut it from the mother plant:

- **First, air layering saves space.** It's the perfect method to use on large-and-in-charge aroids such as *Monstera deliciosa*. Because you're propagating the plant while it remains intact, you don't have to make room for a giant propagation box or a large vase to hold a huge cutting.

- **Second, air layering simplifies your plant care routine.** With air layering, you only have one plant to take care of: the mother plant. This is because the future baby is still attached to the mother plant, much like how a human baby remains attached to the mom throughout pregnancy until birth when the umbilical cord is cut. Likewise, your future plant baby will continue to receive moisture and nutrients from the parent plant as it grows roots—until you snip it! You're welcome for that umbilical cord visual the next time you take a stem cutting.

- **Third, air layering increases your chances of winning the race against rot!** Because the baby remains attached until you cut it off, there is less risk of that section rotting than if you cut it off first. This is why air layering is beneficial for variegated aroids that are susceptible to rotting during other methods of propagation. If roots do not grow during air layering and the mother plant is still healthy, you can simply remove the substrate and the plant will continue to grow as if nothing happened.

- **Finally, air layering tends to produce larger new plants faster.** Due to the constant flow of moisture and nutrients your future cutting receives from the mother plant's roots, it will be more robust than if it had to fend for itself as a cutting while trying to grow its own roots.

On the other hand, the one reason why I don't like this method of propagation is that I don't think it's aesthetically pleasing. It drives me bonkers to see a plastic baggie filled with soggy moss attached to my beautiful plant's stem. However, if it's a plant that I want to propagate most efficiently, I am willing to make the sacrifice.

My too-tall philodendron 'Pink Princess'.

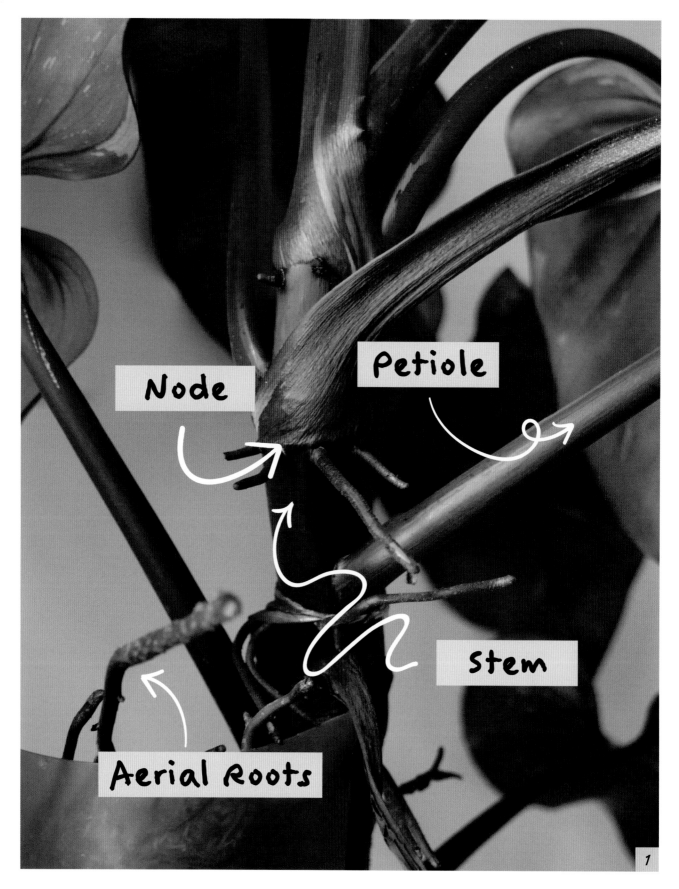

Node

Petiole

Stem

Aerial Roots

1

THE ULTIMATE GUIDE TO HOUSEPLANT PROPAGATION

Ideal For

- Large climbing philodendrons (e.g., *Philodendron erubescens* 'White Princess', *Philodendron sodiroi*)

- Monsteras

- Variegated vining aroids (e.g., variegated syngonium [*Syngonium podophyllum* var. *albo-variegatum*])

- Ficus* (e.g., fiddle leaf fig [*Ficus lyrata*] and rubber tree [*Ficus elastica*])

*Air layering a ficus will look slightly different than described here and involves first removing or wounding the outer layer of bark in the area where you place the rooting sack. It is much easier to propagate ficus by taking a simple stem cutting.

Supplies Needed

- Rooting hormone gel or powder (optional)
- Sphagnum moss
- Clear plastic wrap (alternatives: propagation ball, small plastic cup cut on the sides)
- Gardening hook and loop fastener or string
- Sterilized cutting tool
- A new pot for your prop
- Aroid mix

Timeline

- 3–5 weeks to fully rooted

Instructions

Step 1: Know thy Nodes!

Decide which node you'd like to be the base of your new plant. From that node, adventitious roots will grow. That's the node you're going to wrap up.

Step 2: Prep the Sack

You need three things to make a proper propagation sack for your roots to grow in: substrate + sack + ties:

- For the substrate I prefer to use moist sphagnum moss because it's less messy than aroid mix or perlite in this situation and retains moisture like a sponge, but I encourage you to experiment.

- Use plastic wrap, a little plastic cup, or a rooting ball to hold the moss against the node (the sack).

- Use string or a gardening hook and loop fastener for the tie.

Step 3: Bag the Sphag

Smoosh moist sphagnum moss around the node in a big spongey blob and then bag it up in plastic wrap. The plastic wrap serves to keep the moss together AND to prevent it from drying out. Secure the plastic with your tie on top and on bottom. Leave a tiny hole or flap in the plastic where you can add water when it's time to remoisten the moss. Check on your propagation sack at least once a week. If you notice the moss starting to look or feel dry, water it liberally.

Step 4: Clean the Roots

In approximately 3 to 4 weeks, once the roots are 1½ to 2 inches (4 to 5 cm) long, you can remove the plastic wrap. Clean at least 80 percent of the moss from the roots if you plan on potting the new plant in a chunky aroid mix (most common). Alternatively, remove all the moss from the roots if you'll be potting it up in LECA or Lechuza pon in a self-watering pot.

Step 5: Take a Rooted Top Cut!

It's time for that plant kid to stop mooching off its mom. Grab the sterilized cutting shears and cut along the stem about 1¼ to 2 inches (3.2 to 5 cm) *below* the newly formed roots to reveal your new independent plant babe, a.k.a. the top cut.

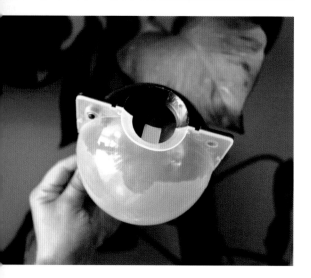

As mentioned in Step 2, these handy rooting balls are a great reuseable alternative to plastic wrap.

SIMPLE LAYERING

Featuring *Hoya heuschkeliana* var. *variegata* in potting mix

Simple layering is just that: simple. While not as popular as other propagation methods, simple layering can be a fun and effective way to propagate vining/trailing houseplants. Just as you are more successful with your support system right there by your side, your propagules root faster while still attached to the mother plant, receiving a continuous supply of water and nutrients.

Ideal For

- Vining houseplants, such as:
    - *Hoya carnosa* (wax plant)
    - Spider plant (*Chlorophytum comosum*)
    - Pothos (*Epipremnum aureum*)
    - Heartleaf philodendron (*Philodendron hederaceum*)
    - Tradescantia species

Supplies Needed

- Sterilized cutting tool
- A second pot filled with substrate (peat moss and perlite mixed 50/50 works just fine here, as does sphagnum moss or go straight to a well-draining potting mix such as aroid mix, which works well for hoyas)
- Floral pins, bobby pins, or opened paper clips

Timeline

- 6–12 weeks to fully rooted depending on plant species and environment. Note that all nodes may not root.

Tip

Because this is not often done using a humidity enclosure, such as a propagation box, it's ideal to wait until spring or summer when it's warm and humid in or outside your home and your plant is actively growing. Otherwise, you can speed things up by doing this any time of year inside of a grow tent or cabinet, or by using a humidifier.

1A 1B

Step 1: Setup
Choose the plant you want to propagate and place a smaller pot beside it, filled with moist propagation substrate or potting mix.

Step 2: Lay Down the Vines
Lay the vines you want to propagate across the top of the second pot. Remove any leaves around the nodes you will be burying into the substrate, especially if they are in the way. I didn't need to remove any on this plant.

2

3

4

5

Step 3: Pin Down the Nodes

Bury each node along the stem of the vine into the substrate. Use opened paper clips, floral pins, or bobby pins to hold them down if needed.

Step 4: Keep the Second Pot Moist Throughout the Propagation Period

Water the mother plant as you normally would, but keep the second pot evenly moist throughout the entire propagation period. A high humidity environment helps with this.

Step 5: Separate the Vine

After 6 to 12 weeks (or longer, depending on the plant species), a new adequate root system of at least 1 inch (3 cm) in length will have formed on at least some of the nodes of the vine layering in the new pot. When it does, you can cut the vine free from the mother plant.

Step 6: Leave Propagation in Place or Repot

After freeing the rooted vine from the mother plant, you can leave it as is (since you layered it into a new pot) or cut the vine into individual rooted nodes and pot each piece up separately to make multiple new plants.

If your vine rooted into a temporary substrate for propagation purposes, transfer your rooted vine or the individual rooted nodes into the potting mix or substrate of your choice.

Step 7: Aftercare

After the roots are roughly 1 inch (3 cm) long, begin to water and fertilize as you would the mother plant, letting it dry out completely between waterings. Hoyas love bright, indirect light and even enjoy a few hours of direct sunlight per day. Your new plant is an exact clone of the mama!

My new, handmade, fully rooted, variegated *Hoya heuschkeliana* plant!

SEXUAL PROPAGATION AND POLLINATION

Featuring Anthurium species

*How did the anthurium in male anthesis
hook up with the anthurium in female anthesis?*

He inflo'ed into her DMs

If pets are the new kids and plants are the new pets, anthurium breeding is the new puppy breeding. Interest in anthurium pollination has increased tremendously in recent years due to the skyrocketing popularity of houseplants and, specifically, aroid cultivation.

Anthurium pollination is much more efficient than propagating anthurium from stem cuttings. Anthurium collectors propagate anthuriums from stem cuttings all the time, and it's common to see butt cuts for sale on social media. However, it's not as easy to regrow a finicky anthurium from stem cutting as it is to propagate an eager-to-please philodendron. The risk of root and stem rot is much higher. The price point of the plant you are working with is often higher as well, as my credit card statement will tell you. In the same amount of time you could grow a few plants from cutting up the stem of your anthurium into little pieces (which also puts your mother plant at risk for crashing if not done correctly), with pollination, you could end up with five, ten, even one hundred seeds with little to no stress on the mother plant.

How many babies result from pollination depends on how many flowers are successfully fertilized on the inflorescence. Older plants generally have longer spadices containing more flowers than younger plants of the same species, providing more opportunities for berries and seeds to form. If you have not already, I encourage you to read chapter 2 for an in-depth overview of pollination and fertilization.

This all being said, if you want to make an exact clone of your anthurium, taking a stem cutting or dividing off a pup it produces (if you are super lucky) are wonderful ways to go. While the propagation of anthurium through stem cuttings or division only yields a few offspring, the new plants will be genetic clones of the mother plant and will maintain the mother plant's vigor. Therefore, while it's not the most productive form of anthurium propagation, asexual propagation of anthuriums certainly has its place.

Pollination is hands down my favorite form of propagation, and anthuriums are my favorite plant species to do it with. I consider ornamental plant breeding a form of living art, and the inflorescences are a blank canvas onto which I paint pollen. My final creations are the many beautiful and colorful hybrid plants that grow because of my pollination efforts. Every day I try to push the boundaries of nature and science to achieve new levels of botanical beauty.

Anyone can walk into a garden center and buy a new plant, but creating a new plant through breeding two of your existing plants is an art form reserved for only the most curious plant lovers. People who have a desire to create plant hybrids are not merely plant collectors; they are the inventive mad-scientist types among us. They are the tinkerers and the explorers. Just by having an initial interest in this section of the book, I can tell you fit the bill.

Who wouldn't want to create more anthuriums?
Pictured: My *Anthurium antolakii* mother plant.

Anthurium Infructescence
(a fertilized inflorescence)

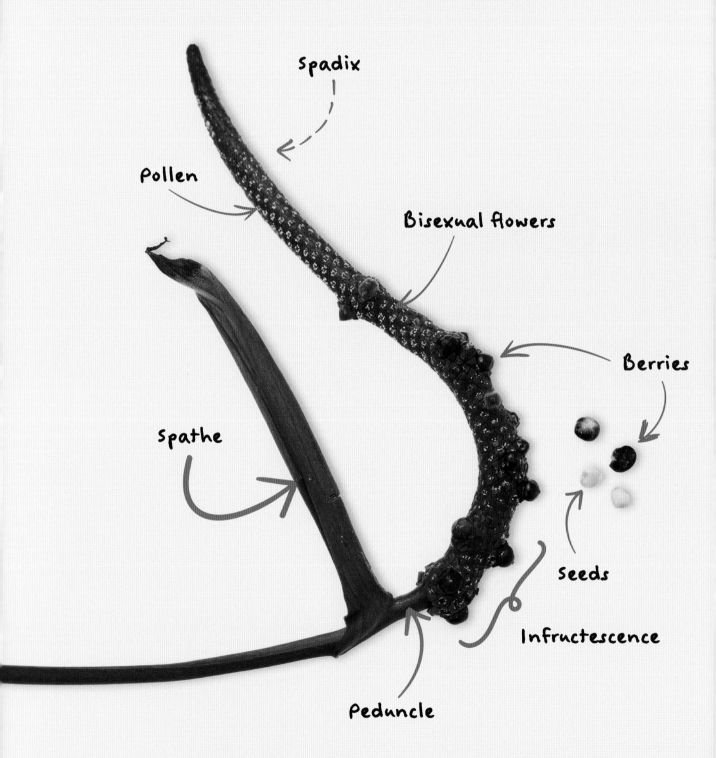

Spadix

Pollen

Bisexual flowers

Berries

Spathe

Seeds

Infructescence

Peduncle

ANTHURIUM REPRODUCTIVE ANATOMY

As with other aroids, the flowers on an aroid inflorescence are actually the near-microscopic bumps covering the spadix. Each one of the flowers on an anthurium spadix contain both male and female sex organs—stamens and pistils—and are therefore considered bisexual. Once the flowers are successfully pollinated and fertilized, the inflorescence is called an *infructescence* because it is bearing, or will soon bear, fruit. The technical definition of infructescence is "an ensemble of fruit" and I think this is an even better way to describe the group of colorful, happy little berries that form on your anthurium after pollination.

ANTHURIUM SEXUAL SYSTEM REVIEW

As discussed on page 36, to avoid self-pollination and promote outcrossing (breeding with other plants), most anthuriums' bisexual flowers first go through a receptive female stage, called female anthesis, and then a male stage, called male anthesis, where it presents pollen.

Let's imagine this as a dramatic short story taking place on a single anthurium inflorescence in a dark and rainy forest in Panama:

All the flowers on the inflorescence begin in female anthesis. They are dripping with stigmatic fluid. One calls out, "Hey guys, we ladies are ready to receive pollen!" *(Work with me here, I'm trying to make this as rated-PG as possible.)*

However, no other flowers on the spadix answer because they're all also in female anthesis at the moment. All you can hear are crickets. Literally, because you're in the forest.

Time goes on, and the lonely female flowers dry up.

One by one, starting at the bottom of the spadix, the flowers start to transition into male anthesis, build their mojo-dojo casa houses, started calling each other bro, and producing pollen. "We are male now!" they shout. "Who wants some pollen? Any ladies around here want to get pollinated?"

But they are *all* dudes now, so no one can receive pollen and put the sperm to good use.

It's a love story almost as tragic as Romeo and Juliet. And it helps to prevent the plant from pollinating itself. . . . Unless of course when the flowers were first in female anthesis, you or a natural pollinator took pollen from *another* inflorescence with flowers in male anthesis (or pollen previously collected and stored) and applied it!

And this, dear reader, is how you time the pollination of anthuriums!

The timing of the inflorescence's reproductive stages is variable, and the entire sequence can range from 1 week to over 30 days to complete. The amount of time it takes from start to finish tends to increase with the number of flowers per inflorescence. The longer the inflorescence, the more likely you will have a longer flower cycle.

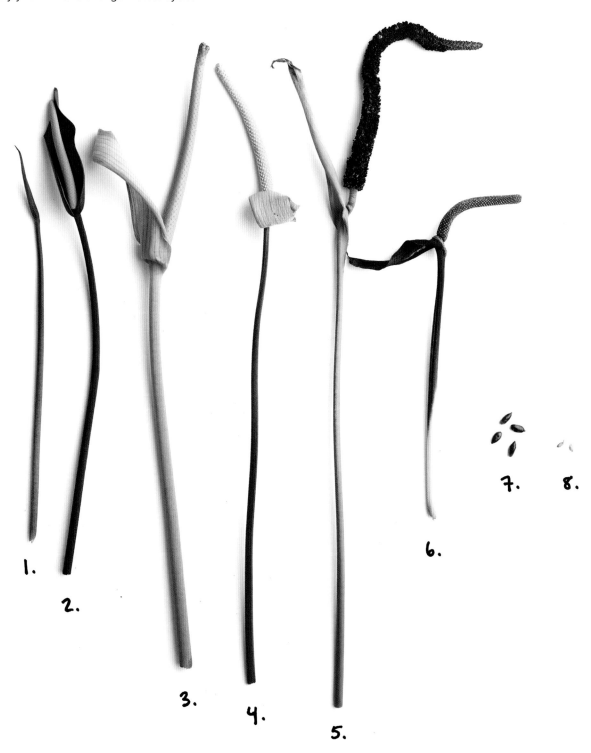

1.

2.

3.

4.

5.

6.

7.   8.

REPRODUCTIVE STAGES OF ANTHURIUM INFLORESCENCES

The following are the reproductive stages of an anthurium inflorescence:

(Note these are flower stalks from different species of anthuriums.)

Stage 1: Are You Happy to See Me?
An inflorescence emerges. Here you see a long peduncle with a small spathe. It's a grower, not a shower. The peduncle, spadices, and spathes of anthurium species are all different sizes, colors, and shapes.

Stage 2: The Big Reveal
The spathe opens revealing the spadix.

Stage 3: It's a Girl!
Female anthesis begins and droplets of stigmatic fluid run up and down the flowers on the spadix. Some anthuriums' flowers will all produce fluid simultaneously, and other species will produce it in succession. Some anthuriums produce an excessive amount of fluid and even drip onto your table, and others produce so little fluid you can only see it with a magnifying glass. The time range for this stage is incredibly variable and depends on the species.

Stage 4: Now It's a Boy!
Male anthesis begins. This stage lasts two to three times longer than the female phase.

Now that the stigmatic fluid dried up, pollen begins to form on flowers at the base of the spadix and works its way up over time.

Stage 5: What the Fruc?
If the flowers were pollinated and fertilized, your fruit ensemble forms! Congratulations, you have an infructescence. Most of the berries on this one were picked off already, but you can see a few small ones remaining up top.

Stage 6: RIP
This is what is a dying infructescence looks like. All good things must come to an end.

Stage 7: Anthurium Berries
Fruit of the plant collector gods. Also, bird snacks!

Stage 8: Anthurium Seeds
Seeds shown here were removed from inside the berries. They will be planted to grow new anthuriums that will then produce more inflorescences.

Fun Fact

A few species of anthuriums do self-pollinate as a way of survival. Anthuriums such as *Anthurium scandens*, *Anthurium gracile*, and *Anthurium bakeri* are considered **apomictic**, meaning they are self-pollinating and can form seeds and berries without outside help.

ANTHURIUM POLLINATION PREPARATION

Because the bisexual flowers on your anthurium's inflorescence will be in female receptive mode first, as a grower, you need to have (male) pollen on hand from another inflorescence when your flowers are ready. Never keep a woman waiting, and that goes for anthurium flowers too. There is no doubt that the most viable pollen to use for pollination is fresh pollen. If you can use fresh pollen off another flower currently in male anthesis either from another inflorescence on the same plant or, more likely, off another plant you have nearby, that is the way to go. However, unless you are growing many mature anthuriums at once, this is often not possible.

Alternatively, you could have a friend with anthurium pollen nearby on speed dial like I did the first time one of my anthuriums bloomed. When I texted my friend, Amanda (@_Bunnyyy___ on Instagram), she drove over right away, stopped in front of my house, and rolled down her window. She reached out her hand and handed me a small piece of folded foil, wished me good luck, rolled up her window, and drove away. I received quite a few curious looks from my neighbors who were out in their yards as it looked like there was something other than pollen being exchanged. It was worth the suspicious looks though because pollination was successful.

The most common and efficient scenario, however, is that you build up an anthurium sperm bank in your freezer. If you don't have one started, this section will help you start one. This way, when the moment arises and another inflorescence is ready, you'll have a variety of donors to choose from. It will also intrigue friends and family when they go to look for ice cream in your freezer and you tell them what it's located next to!

Collecting and Storing Pollen

Every plant hybridizer has their own method for collecting and storing pollen they swear by. There's truly no one correct way to do it. However, if you follow the guidelines below, your anthurium pollen should remain viable for 3 months to 1 year. This advice is applicable to all aroid pollen, although the amount of time the pollen you collect remains viable in storage depends on the genus you are working with.

Pollen Collection

- **Keep it contained:** I like to store pollen in a piece of foil folded up because its wide surface area makes catching every grain of pollen you brush off easy and its waterproof. Alternatively, people collect pollen into glass or plastic vials or onto black paper. I keep all my folded pieces of foil in a plastic zipper-top bag, but you could just as easily use any container that fits in your freezer.

- **Keep it cold:** Place the packaged pollen in the freezer and don't let it thaw. If you have a power outage, or in the case of a natural disaster or alien invasion, do not forget to run and rescue your pollen.

- **Keep it dry:** Wet pollen is nonviable pollen. Use the silica gel bead desiccant sachets you find in vitamins and new shoe boxes in your plastic bags to absorb moisture. You can also buy the beads online.

- **Label it:** Use a permanent marker to label the outside of your vial or envelope. Describe the type of pollen and date.

NOW IT'S YOUR TURN!

[left] My pollen storage bags (a.k.a. anthurium sperm bank). Remember, each grain of pollen contains two sperm.

[page right] Use a paintbrush to brush pollen off the inflorescence onto a piece of foil, paper, or into a vial. If you brush and nothing comes off, pollen is not available. It's important to note that some inflorescences skip the pollen-producing stage altogether, especially if it's on a less mature plant. If this happens, you will have to wait for another time. Don't scrape the flowers too hard or you risk damaging them.

ANTHURIUM POLLINATION STEP BY STEP

Timeline

- 4–12-plus months from pollination to seed production depending on the anthurium species involved*

*For example, the popular *Anthurium clarinervium*, *A. veitchii*, *A. besseae aff*, and *A. luxurians* seed parents will be on the longer end of that spectrum.

Ideal For

- All species of anthuriums

Supplies Needed

- Viable anthurium pollen
- A clean paintbrush or clean fingers
- In 4–12-plus months:
    - Strainer to rinse berries and small jar to soak
    - Seed tray or propagation tray with clear lid to plant seeds in
    - Seed planting substrate

Notes

- Here I am using a paintbrush, but typically I just use my fingers.
- Also, the photos shown in this section are not all from the same anthurium plant.

Instructions

Step 1: Load Up on Pollen

Grab pollen stored from the freezer or gather fresh pollen. Here I am collecting fresh pollen from my anthurium 'Red Crystallinum' hybrid, in male anthesis. Don't be shy. Get the pollen all over your brush like a pig rolling in mud . . . or a fly buzzing all over an inflo! To help pollen adhere better to the brush, wet the brush first by dipping it in the stigmatic fluid on the inflorescence in female anthesis.

Step 2: Deposit Pollen on Receptive Inflo

Find the receptive inflorescence in female anthesis on the anthurium you want to pollinate. In some species, receptive flowers will be dripping wet with stigmatic fluid, and on others, you will barely be able to see that its receptive at all. In species where fluid is difficult to see, I use a magnifying glass with a light to check for slight moisture to see if the inflorescence is ready to be pollinated. You also can smell it as they will typically emit a scent when ready for pollination.

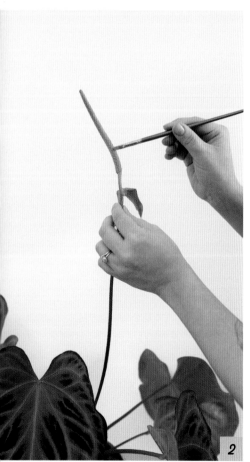

Step 3: Watch Berries Begin to Form

I use my *Anthurium antolakii* as an example of a fertilized anthurium from this step onward.

"Did it take? Did the pollen take?" This is what you will wonder every day as you sit and stare at your inflo. Staring super hard at it helps fertilization to occur, I hear. Raised bumps along the spadix will be your first sign that the pollen took.

Step 4: Let the Magic Happen

If berries form on your inflorescence, it is now an infructescence. Berries may be green at first and then turn to orange or red depending on the species, as they ripen—similar to other fruits you are familiar with.

If the flower stalk itself starts to turn yellow, brown, or multiple psychedelic colors at once (yellow, orange, green, red), that is not a good sign. That means the plant is likely aborting fruit production. Once this begins to happen, there is nothing you can do but make plans to start over again another time. It's okay, this is nature!

Anthurium berries can take 4 to 12 months to ripen on an inflo?

You've got to be pollen my leg!

Step 5: Berry Picking Season!

Four months to over a year later, depending on the anthurium species and your environmental conditions, it will be seed-harvesting time. You will know the berries are ready to pick when they are both dark in color (typically a shade of red) and start to pop out of the infructescence a bit. There is often a band of white at the base of the berry, depending on the species, as you see here. Some berries may even hang from the flower stalk from two thin strands. They remind me of extremely loose children's teeth! On some anthurium species, the berries won't protrude as much, but they darken and feel soft to the touch when it's time to pick them.

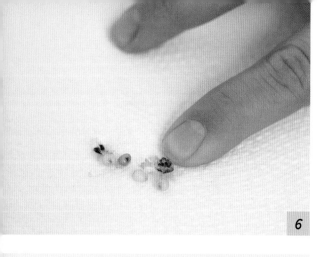

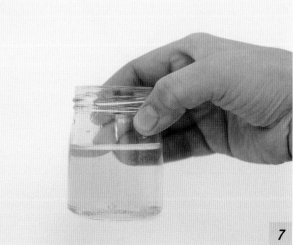

Step 6: Extract the Seeds

Using a paper towel, reusable napkin, or similar cloth as your working surface, roll each berry with the tip of your finger while applying downward pressure to remove the seed(s). Smoosh the berries, not the seeds. Each berry contains one to two seeds.

Step 7: Clean 'Em

Lucky for you sensory seekers out there, anthurium seeds are coated in a gooey, gelatinous coating called the **mesocarp**. Gently rolling the seed along your cloth should remove most of it. To help remove even more, place the berries in a strainer and rinse them off. Alternatively, place the seeds in a jar of water, and let the seeds soak for a day or two. Swish the seeds around and strain out any outer coating that naturally floats off the seeds. If you leave the seeds soaking for multiple days, change the water daily to keep it fresh. I'm not going to lie. I've definitely left seeds soaking for a couple weeks before (changing the water daily), and they started to germinate and grew just fine. However, this is not the ideal method for growing anthuriums.

Every seed develops at its own pace. Some seeds will already have a tiny green root growing out of them. This tiny initial root is called the radicle and its early appearance means the seed is an overachiever. It is already starting to germinate. Some seeds may be white, some seeds may be green. The color depends on the species of anthurium and also how long you've waited to pick the berries. The longer you wait, the greater chance they will start to turn green while still inside the berry.

Step 8: Plant the Seeds

I have conducted numerous experiments with hundreds of anthurium seeds. While the results varied slightly each time, they were significant enough for me to conclude that planting in Fluval Stratum results in the fastest and most robust growth. Second to that, I recommend moist sphagnum moss alone or a moist sphagnum moss layer on top of aroid mix.

Fill a seedling tray or container that has a clear lid with substrate. Then, place each seed right on top of the substrate of your choice if you're looking to save time. However, even better would be to first create a little divot in the substrate with either a seed dibber or by pushing the seed down gently with your finger. The seed then will be surrounded by the substrate on all sides, except for the very top. This maximizes the surface area of the seed that is touching the substrate, while also giving it easy access to grow upward. Planting in this way helps prevent the seed from moving around when you water.

Step 9: Seed Germination

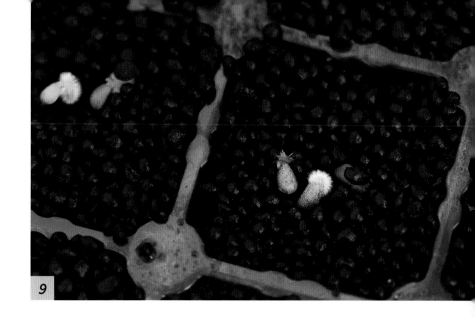

9

Anthurium Seed Lifecycle

1. Germination typically takes 1 to 3 weeks to occur after planting, depending on the seed.
2. Once it germinates, the fuzzy little **radicle** will pop out. Boop!
3. The radicle will grow into one or two white terrestrial roots. While the roots grow downward, the first leaf will grow, and from that leaf, another slightly larger one!
4. The roots grow longer and leaves grow larger and eventually a third leaf emerges.

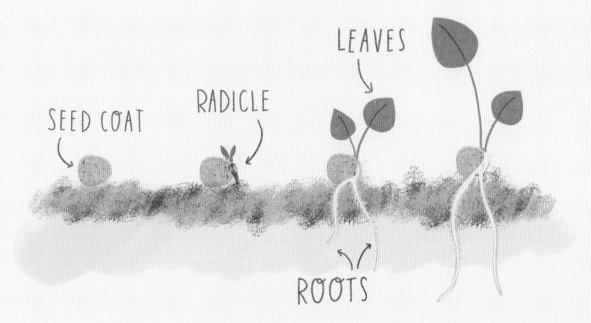

SEED COAT   RADICLE   LEAVES   ROOTS

1.   2.   3.   4.

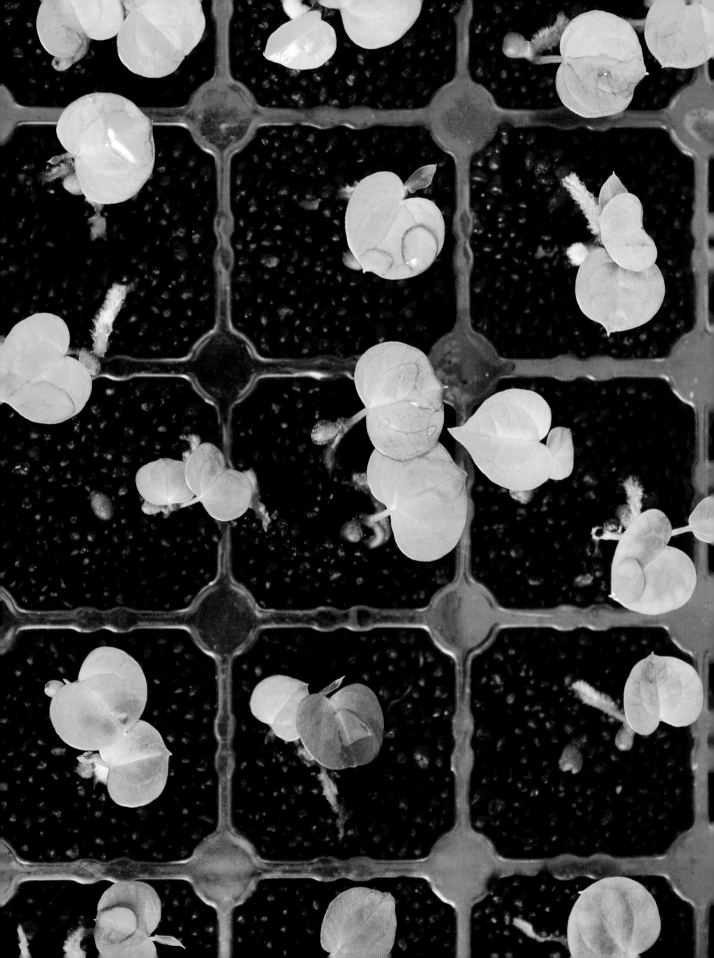

Step 10: Transfer to Mix

It is at the two-leaf seedling stage that you ideally pot your seedlings into anthurium mix or pon (or whatever substrate you intend to grow them in). I like to wait until the two leaves are fully grown to help it withstand the shock of transplanting. Of course, you can transplant them earlier, or wait longer, but I find that around this time is the sweet spot. If you transplant them when they are too little, they may not do well with the shock of a new substrate and changes in humidity. If you wait too long, and seeds are still in a covered seedling tray, you risk the spread of fungal or bacterial infections from too much condensation and moisture, no airflow, and overcrowding inside your tray. Also, the roots of the seedlings will eventually outgrow the tray affecting the health and vigor of the seedlings.

It is also at this two-leaf stage that it is safe to fertilize the seedlings with diluted fertilizer. Most fertilizers have recommendations on the back for what dosage to use with seedlings. If not, a good rule of thumb is to dilute the fertilizer to one-fourth to half the amount you would use for a general feeding to avoid over fertilizing.

Pot your seedling up in a small 3 to 4-inch (8 to 10 cm) pot with drainage holes if using a chunky anthurium mix, or many anthurium growers use pon in a semi-hydroponic setup. For at least the first 3 weeks after transplanting, keep the plants in the highest humidity possible; 75 to 100 percent is ideal. Water them as you would any of your mature anthuriums: thoroughly and never letting the substrate dry out.

Step 11: Aftercare

After the adjustment period, acclimate the plants to more typical growing conditions. Like all your other anthuriums, never let the substrate dry out completely, and grow in indirect light. Once they begin to grow four leaves, it's safe to fertilize using your nutrients at regular strength according to the instructions on the label. Many serious growers additionally adjust the pH of their water since anthuriums absorb the most nutrients when their water is slightly acidic, in the pH range of 5.5 to 6.5.

More information about growing anthuriums can be found on my Instagram @alltheplantbabies and blog at www.alltheplantbabies.com.

Troubleshooting

There are a few reasons why pollen doesn't take or why flowers abort.

1. **Environment:** One reason could be that your environmental conditions were not ideal for successful pollination. Did the surrounding temperature drop below 65/70°F (18/20°C) for an extended period of time? Was it excessively hot (over 100°F) (40°C)? Was it too dry (less than 40 percent RH)?

2. **Pollen viability:** Perhaps it was not due to environmental conditions, but instead your pollen was no longer viable or fresh. How long pollen remains viable depends on the species of plant, as well as the environment conditions in which you store the pollen. The longevity of pollen can range from a few minutes to up to a year. Ideally, you transfer pollen from one infloresence to another without having to store it. However, most people don't have enough plants flowering at the same time for this to work out and they must store pollen to wait until another infloresence is in the female stage. Keeping pollen dry and freezing it extends its life. On average, pollen from the anthurium species I work with stays viable for up to 6 to 9 months kept dry in the freezer inside foil within a plastic bag.

   If the flowers on your infloresence abort, ask yourself: How did you store your pollen? Was it kept dry the entire time? How old was it? Unfortunately, the expiration date doesn't come printed on the infloresence so you will have to observe and keep notes. Sometimes, our failures are our biggest lessons.

3. **Genetic compatibility:** A third reason is that the two anthuriums you crossed may not have been genetically compatible. With over 950 species recognized and over 2,000 expected to exist, *Anthurium* is by far the largest genus in the aroid family. Over the years, botanists and taxonomists have partitioned anthuriums into different groups called sections based on their structural characteristics. These sections help to suggest which anthurium species will be easiest to cross with which, and while cross-sectional hybrids are not unheard of, they do not happen successfully very often.

As a quick review from chapter 3, all angiosperms (of which anthuriums are a part) are amazing. When you place a pollen grain (the male part) on a stigma (the female part of the flower), what happens next is like the screening of someone's online dating profile: there are a few critical nonnegotiables that must be asked and answered first before engaging further.

The stigma scans first to see if the pollen grain is from the same species or a compatible one, and/or if the grain is from the same plant.

Why is compatibility so importation? Sometimes pollen from an incompatible plant is just not capable of growing a pollination tube long enough to reach the ovary. It is common for anthurium breeders to find difficulty in breeding anthuriums across the sections in which they are taxonomically divided. This could be because, as observed in other angiosperms, sometimes the pollen tube gets part of the way down the style, then turns around and tries to grow back up toward the stigma in closely related species that are considered incongruent. There is just too much evolutionary distance between them. When pollen does take to create an intersectional cross, either the inflorescence is aborted or, the seed yield will be extremely low.

4. **Bad vibes:** Last but not least, the vibes could be off. Is Mercury in retrograde? Are your chakras out of balance? I'm sorry, but the instructions for fixing these issues are in a different book. Perhaps you need to take a break and head to page 190.

TISSUE CULTURE

I bet you never imagined that most of the houseplants you buy from the store start off as a cluster of cells in a laboratory, but here we are. Everything from a grocery store succulent to a big box store peace lily to the variegated monstera in a fancy plant shop most likely started off as plant tissue in a tissue culture lab.

Tissue culture, or micropropagation, was developed in the 1960s to asexually propagate plants (create clones) in a laboratory. Using just a small amount of plant tissue in vitro in a sterile environment, vigorous, disease-free plants can be created in large numbers. This technique revolutionized horticulture, agriculture, and plant research. Now, plants can be cloned and sold by the hundreds or thousands, year-round. This level of mass production is why most houseplants are affordable. A laboratory only needs a few mother plants to produce plants by the hundreds for wholesale.

Tissue culture is also the reason why it is challenging to have a small business selling plants created only from vegetative cuttings or storage structures such as alocasia corms. It takes months for you to fully propagate a single plant at home, but tissue culture labs can propagate hundreds in a flash. And, with the tremendous volume of plants they are producing, their selling price will be lower than plants propagated by hand.

Therefore, some plant shops buy plants wholesale from tissue culture laboratories, acclimate them, and then sell them to customers. Others, especially larger online plant shops, are getting into the tissue culture business themselves and starting their own laboratories to continue to remain ahead of the competition.

A new trend, reserved for the most ardent of houseplant enthusiasts and sellers, involves creating at-home tissue culture laboratories. You can pay to attend trainings to learn the process. You must create a sterile environment in a section of your home and purchase the correct equipment. But after that, just think of the wild propagation stories you'll be able to tell at parties!

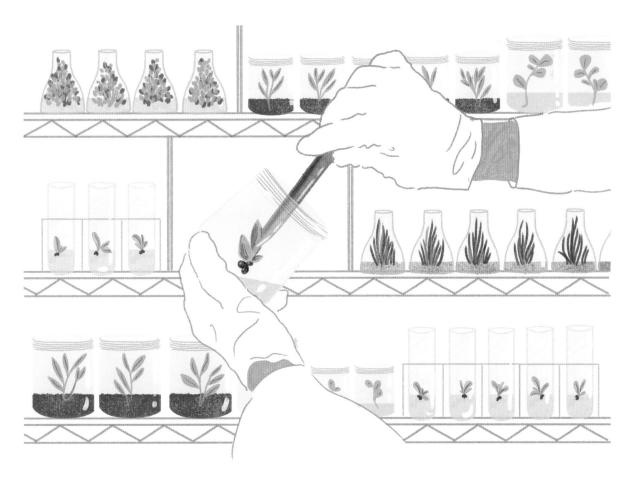

SUCCULENT PROPAGATION

You're still reading this book? You must be a real succa!

I know what you're thinking. Lindsay is a tropical plant collector; who dis?

But I want you to know I turned over a new leaf. A succulent leaf. And it had roots and a baby attached and now I'm hooked. The wish list has begun . . . somebody stop me!

I won't go it alone on this journey, however, and so I'm taking you with me. I want you to also experience just how much fun it is to propagate these plants, even if you, too, think of yourself as more of a tropical plant person, or maybe even not much of a plant person at all. Or perhaps you are already a succulent expert and are here for the corny jokes?

A succulent is any plant that has thick, fleshy tissues that can store water. This includes cacti. If each plant species had its own personality, succulents would be independent and self-sufficient but would require being in the spotlight (sunlight) at all times. Their babies would be the same. You don't want to overparent these resourceful kids. Succulent propagules—the leaves you pull off, the stems you cut, and the offsets you remove—already have everything they need to produce new roots and leaves inside of them. They even have their own water supply. All you need to do is place them in a healthy environment and *not* smother them with your love. No babying or helicopter plant-parenting warranted!

Succulents are so much fun to propagate and yield so many babies that once you know how to do it, it's hard to own one and purposefully keep it intact. Not only can each individual leaf on the plant produce an entirely new plant, but new growth often emerges from a remaining healthy rooted stem. They are the gift that keeps on giving and giving.

What gives succulents the magical superpower-like ability to produce completely new growth from different plant parts is undifferentiated **meristem tissue**. This tissue contains cells that can divide continuously to form plant parts such as new leaves and roots. Meristem tissue is found on areas of the plant called (you guessed it) **meristems**, which, when located on the tips of roots and shoots, are called apical meristems. When located elsewhere on a plant, it is called secondary meristem tissue. Meristems are responsible for the primary growth of any plant, not only succulents.

On most succulent leaves, secondary meristem tissue with totipotent cells is located on the base of the leaf. This is why when a succulent leaf is removed, it has the potential to grow an entirely new plant from its base. Whether this happens or not depends on if the meristem tissue remains dormant or becomes active due to environmental conditions. Meristem tissue also is found on the stem, which is why after beheading most succulents, you will notice it will begin to produce new growth again from the main stem.

Like tropicals, if you are growing succulents indoors in a temperature-controlled environment with the support of grow lights, you can propagate with great success year-round. Otherwise, it is ideal to wait until the warmer and sunnier days of spring and summer that bring about active growth to perform these techniques.

There is meristem tissue on the bottom of each of these succulents' stems that is capable of producing new roots and on the top of each is tissue capable of producing new leaves.

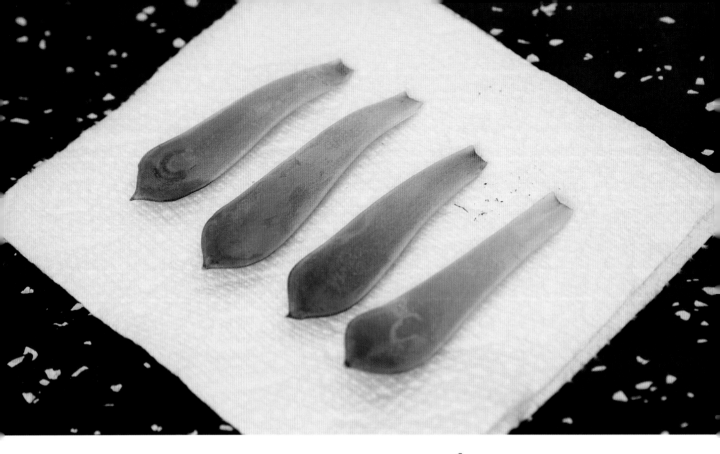

SUCCULENT LEAF PROPAGATION

The most common and intuitive form of succulent propagation is leaf propagation. It's the first type of plant propagation that many people try. If a succulent leaf falls off onto your shelf or the floor, it will most likely grow new roots and a little plant right there where it landed. It's how a succulent would reproduce if an animal or strong gust of wind knocked off a leaf in the wild.

Since some succulents produce an abundance of leaves, this is a great method to produce many new plants at once. However, keep in mind that it will take longer for a full plant to grow from a small leaf than it would from a stem propagation. Succulent leaves can root in several ways, the most popular being left out dry (also known as air propagation), in moist soil, and water propagation. We will review the dry and soil propagation methods in the following section since they are the most common and yield the best results.

[above] Succulent leaves callusing on my propagation mat

[page right] My Variegated Ghost Plant (*Graptopetalum paraguayense* 'Variegatum') that I propagated and grew from mature leaves.

DRY SUCCULENT LEAF PROPAGATION

Featuring *Echeveria pulidonis*

The number one reason why people screw up succulent propagation is overwatering. If this has happened to you—or sounds like it *could* happen to you—give this method a try. This is a hands-off method of succulent leaf propagation, and when I say hands off, I mean don't even THINK about putting your hands on the mist bottle or watering can! This method is ridiculously easy but may be a lesson in self-restraint if you are naturally an overnurturer.

This method is also ideal for new succulent parents as there is not any watering guesswork involved. However, because you won't be watering or misting the succulents or the soil at all, this method will work best when there is a little bit of humidity in the air. A relative humidity of 40 to 70 percent is ideal for this method.

Ideal For
- Succulents that have thick leaves such as echeveria, graptoveria, Jade plants, crassula, and sempervivum

Timeline
- 1–2 months

Supplies Needed
- A place to lay your succulent leaves out during propagation
- The pots you'll want to plant your new baby succulents in
- Succulent mix

Preparation
Water the succulent you want to propagate 2 to 3 days before you are ready to begin propagation. This ensures the leaves have enough water stored inside of them to sustain the growth of a new pup without additional water being added during the rooting process. As the leaf pushes out a little plant and new roots, it will draw from this internal water supply!

Instructions

Step 1: Remove Your Leaves

Gently twist and then pull the plump, firm leaves off the succulent you want to propagate. Start with the biggest and most mature leaves from the bottom of the plant to increase your chances of success. Unhealthy leaves are not fit for making babies.

Step 2: Place the Leaves on a Dry Surface

Throw them down or be an artist. I like to make succulent mandalas like you see here, laying them out in a beautiful design. I chose to lay them on well-draining succulent soil in case any of the roots wanted to grow down into it. However, you could just as easily lay them out on a plate or a windowsill to root. It doesn't matter since this is dry propagation—**no water or misting allowed!**

Place the leaves in a place where they receive bright, indirect light, but avoid locations with many hours of direct sunrays to prevent the leaves from burning.

2

Tip

As a former yoga teacher, one of my favorite ways to propagate succulent leaves—either dry or on soil—is to arrange them into a mandala. Mandalas are geometric configurations used by many spiritual traditions for focusing attention and creating a sacred space. The process of arranging these leaves is meditative and relaxing.

3

Step 3: Wait and Observe Growth

After 2 to 4 weeks, the leaves will begin to produce roots. After 1 to 2 months, you will have a baby rosette on most leaves. Toss any leaves that fail to grow a new plant. Don't feel bad—this is just nature! The mandala in this photo was 1 foot (30 cm) beneath two 3-foot-long (91 cm) T8 LED grow light bulbs (the kind that look like shop lights) the entire time and received no water or fertilizer.

Step 4: Prepare the Leaves for Potting Up

Time to transfer your baby succulents to the pots you want to grow them in. If the attached leaf from the mother plant is brown, shriveled, and crispy, that means that the water and nutrient supply for the baby is depleted. Remove it!

If the leaf from the mother plant still has a little squish to it and is yellow, nutrients and water are still inside for the baby. Keep it attached, even as you transfer the baby to its own pot. Eventually the mother leaf will turn completely brown and even fall off on its own if you are patient enough (which I am not).

Step 5: Pot Up the Babies

Plant each new rosette in a pot of succulent soil. Wait another 2 to 3 days and then water the soil.

Step 6: Aftercare

After the initial watering, water once a week. Once fully established, water every 2 weeks to once a month based on the needs of your succulent. Watering needs always depend on your environmental conditions and the time of year. Only water your succulent when it's bone dry. LOTS of bright, indirect light with at least 4 to 5 hours of direct light per day, or bright grow lights, will benefit your echeveria. Make room in those south-facing windows! Fertilize using a succulent fertilizer according to the instructions on the bottle during spring and summer.

4

5

SOIL SUCCULENT LEAF PROPAGATION

Featuring *Echeveria lilacina*

Timeline

- Roots emerging 1.5–2.5 weeks
- Baby plants 2 months

Supplies Needed

- Your willpower not to overmist the leaves
- Spray bottle
- Succulent soil mix
- A pot, container, or saucer

Instructions

Step 1: Remove the Leaves

Select mature, plump, healthy large leaves toward the bottom of the plant. Hold the plant with one hand and with your other gently twist the leaf slightly side to side to loosen it, then pull it off. Repeat for as many leaves as you'd like to propagate.

Tip

Not all your leaves will be successful, so choose a few!

Step 2: Allow Leaves to Callus

Set the leaves aside for 2 to 4 days to allow the end to callus over. Callusing allows the cut end of the leaf to scab over so that when planted, the leaf doesn't start absorbing water and drop dead of rot.

3

5

Step 3: Lay on Soil

Place the leaves on top of well-drained succulent soil in a warm location with plenty of bright, indirect light. A few hours of direct sun each day will likely be tolerated by your leaves indoors as well. If you notice any scorching of the leaves, simply relocate them. If you are working with the long T5 or T8 LED grow light bulbs (like many people hang in greenhouse cabinets and under plant shelves), you can place your propagation 2 feet (61 cm) under one of those for great results. I know because that's how I propagate mine.

Step 4: Mist Sparingly

Mist the soil regularly but sparingly to keep it slightly moist. When you mist, aim for the soil. Try to avoid getting the succulent leaves too wet. If your humidity is over 50 percent, you may be able to skip the misting altogether and your leaves will root just fine.

Step 5: Observe Progress

In roughly 2 weeks, give or take, you should have white or pink succulent roots emerging from some of the leaf tips, and in roughly 2 months, you'll have baby plants. It's okay if some of them did not take root. It's okay if NONE of them took root. Compost the ones that did not or wait a bit longer. In nature, and in plant propagation, only the strongest survive! If after a few weeks, nothing worked, try the dry propagation method.

Step 6: Pot 'em Up

If the original leaf is completely dry and crunchy, you can remove it prior to planting up your pup. Otherwise, pot up your baby with the leaf still attached. The leaf will continue to supply water and nutrients to your baby. Once your pups are planted in the soil, wait for another 2 to 3 days before watering it.

Step 7: Aftercare

After the initial watering, water at least once a week, since they are still small and require more water than mature plants. After a few weeks, definitely move to once-a-week watering. Once fully established, water based on the needs of your succulent given your environmental conditions and time of year. LOTS of bright, indirect light with at least 4 to 5 hours of direct light per day (or bright grow lights) will help your succulents thrive. Fertilize using a succulent fertilizer.

[page left] Removing the original leaf.

Troubleshooting

Six reasons why your leaf propagation failed:

1. You misted it too much and/or you tried this method in too humid of an environment.

2. You selected unhealthy leaves.

3. You didn't have good airflow.

4. There was too much heat or direct light, which burned your leaves.

5. Some succulents with thinner leaves, like aeonium, just don't do as well with leaf propagation.

6. You didn't read this book, you only looked at the pictures.

SUCCULENT STEM PROPAGATION

**Featuring *Pachyphytum oviferum* (moonstones)
and other assorted succulents**

*Why was the succulent so stressed over the responsibilities
that come with being a mother plant?*

She was stretched too thin!

Succulent stem propagation is a simple way to multiply your collection
and propagate succulents with thick stems such as aeonium and jade
plants. It's also an excellent cosmetic fix for rosette–shaped succulents
like echeveria that are *etiolated*, or elongated and pale due to lack of light,
over time. This is a common occurrence when growing rosette succulents
indoors, especially if they aren't placed right inside a very sunny window
or 1 or 2 feet (30 or 61 cm) beneath a grow light.

Once your succulent stretches out, it doesn't go back. Once they stretch
out, they never go stout. So, if your plant looks long and lanky, try this
method out, to get it looking swanky.

For propagation, you can cut a
succulent stem well below the
rosette, directly under the rosette
("beheading"), or somewhere
in the middle of the rosette.

Tip

You can cut a succulent stem at multiple places along the stem. On a rosette-type succulents, this could look like cutting well below the rosette (as shown in this section with the pachyphytum), right under the rosette, so you are "beheading" the succulent, or even slicing the stem right in the middle of the rosette. To slice a succulent stem in the middle of a rosette where it may be difficult to fit scissor blades or a knife, use dental floss or fishing wire to make the cut. Simply wrap a string of dental floss or fishing wire around the stem and crisscross it to behead your succulent.

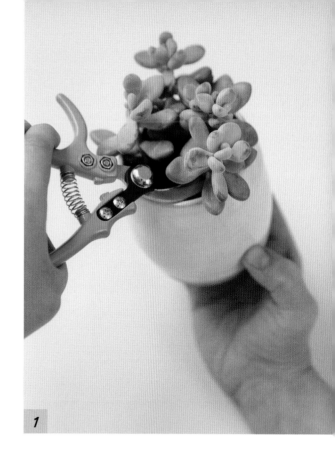

1

Timeline

- 1–4 weeks depending on species, environmental conditions and health of the cutting

Ideal For

- Succulents with long, thick stems, such as aeonium, jade, and kalanchoe
- Etiolated, stretched out rosette-type succulents, such as echeveria, pachyphytum, graptopetalum, and sempervivum

Supplies Needed

- Sterilized cutting shears
- Succulent potting mix
- Rooting hormone optional
- Dental floss or fishing wire if beheading a succulent rosette in the middle of the rosette
- New, small pot for your rooted stem cutting optional (or stick it back in with the bottom cut)

Instructions

Step 1: Cut Your Stems

Using a sterilized, sharp cutting tool (or piece of dental floss or fishing wire), take your succulent stem cuttings. In the photo, I am cutting a little bit below the rosette, leaving a section of bare stem exposed.

2

Step 2: Callus Time!

Let the cut heal and create a seal. Lay cuttings out to callus over for 3 to 4 days to reduce the chance of rot. This is so that when you plant them, they don't absorb too much water and rot. Bonus points if you make a cute design with all your stem cuttings while they callus over.

Step 3: Plant the Stems

Optional first, dip the stem in rooting hormone powder or gel. Then, insert the cut end of the succulent stem into succulent mix so the leaves end up just above the soil line. If working with a rosette, simply lay it right on top of the soil. If you're working with a large number of cuttings, you can stick them as close as an inch or two apart in a large tray. Place them in a spot that gets bright, indirect light. Avoid direct sunrays.

3

Step 4: Mist Carefully

Wait 2 weeks before watering to encourage the stem to produce roots, then mist the soil.

Then, wait 1 more week and mist the soil again. Continue to mist the soil each week for a while, observing your cuttings. If you see your succulent's leaves start to become a little wrinkly, that means the succulent *wants* to be watered—BUT DON'T YOU DARE GIVE IT A GOOD WATERING! This is where I always used to mess up and why I mainly grow thirsty anthuriums now. Your goal here is to make it want water *so bad* that the damn plant starts to produce a whole root system in hopes of finding it.

4

Step 5: Leave Them or Move Them into New Pots

Eventually, the root system will become full and plump, and the roots will be able to absorb water. When you observe a thick root system and the plant displays some resistance when you give it a little tug upward from the substrate, it's likely time you can transfer it to its own pot (or just keep it where it is) and start to water it as you would a mature succulent in your collection.

Step 6: Aftercare

Provide your pachyphytum (or similar succulent) with plenty of bright, indirect light and a few hours of morning or afternoon direct sunlight per day (such as in an east- or west-facing window). South-facing windows can be awesome for succulents as well. Avoid a lot of direct midday sun in hotter climates though, or they may burn. When you water your succulent, water or soak it thoroughly and allow it to completely dry before watering again. Fertilize with succulent fertilizer at least during spring and summer while it's actively growing.

5

LEAF BLADE PROPAGATION

Featuring Snake Plant (*Dracaena trifasciata**)

*No you're not seeing things, this plant was reclassified from Sansevieria.

Native to the desert regions of Africa, snake plants are one of the most resilient, drought, and low-light tolerant large houseplants to own, making them an excellent choice for beginners. These succulents store large amounts of water in their leaves. Therefore, if you are the type of plant parent who forgets to water from time to time, or your home is lacking floor to ceiling south-facing windows, the snake plant may just be the forgiving plant friend you need!

Snake plants also happen to be one of the most fun to propagate. There are very few plants you can just chop in half, stick in water, and wait for new plants to grow, but this one delivers. I'm tempted to end this section right here because it's as simple as that, but I'll provide you will a tad more information because the process is pretty cool to see and the babies are adorable.

Tip

Variegated snake plants are considered **chimeras**, plants that are variegated because of a genetic mutation. This variegation is not stable and can only be passed on to other baby snake plants if you propagate the plant by division (cutting the underground rhizomes). Any babies that grow from a variegated snake plant's leaf blade cutting using the method described here will revert back to the green form. Weird, right?

Timeline
- 2–4 months before you can pot them up

Ideal For
- Snake plant (*Dracaena trifasciata*)

Supplies Needed

All propagation methods:
- Sterilized cutting shears
- Rooting hormone compound, optional
- Succulent potting soil and new pots for the babies

Water propagation:
- Vessel of water

Perlite propagation:
- Plastic container
- Perlite

Soil propagation:
- Plastic container
- Succulent potting soil or propagation mix

Note: Vermiculite, Fluval Stratum, or sphagnum moss also work well for snake plant propagation, but in this section, I will use water and perlite as two examples.

Instructions

Step 1: Cut Leaves

Cut the leaves you want to propagate. You can keep the leaves whole or go wild and cut an individual leaf into 3- to 4-inch (8 to 10 cm) sections. If you cut the leaves into sections, remember that the bottom end is the end that was facing the base of the leaf. Cut the leaf straight across or cut in a V shape to maximize the amount of surface area that will be exposed to the substrate, therefore increasing the area available for baby snake-making. A third option is to cut out the shape of your self-portrait. Please tag me on social media if you do this.

Step 2: Rooting Hormone

Optional, as always: dip the bottom end in rooting hormone compound. Here, I am using rooting hormone gel.

Step 3: Place in Propagation Medium

In the photo, I am using a glass of water. Most people choose to propagate snake plants either in a vessel of water or put them straight into moist potting soil. I also highly recommend perlite propagation (either by simply keeping the perlite moist in a closed-bottom container, or in a semi-hydroponic setup).

Always position the leaves bottom side down into the substrate. Bury the leaf 1 to 2 inches (3 to 5 cm) deep and keep the substrate moist throughout the entire propagation period. Whenever you notice it begin to dry out, give it a good watering or misting.

Alternatively, set up your propagation container hydroponically as described in the tropical plant petiole propagation section on page 88. This way, you won't have to worry about the substrate drying out if there is water in the reservoir container. If you choose to propagate your snake plant cuttings in water, refresh the water weekly, or when you remember, to reduce algae buildup.

A snake mommy

Snake plants propagating in perlite

Snake plants propagating in potting mix

Pilea peperomioides for moral support

Propagator box

4

5

Step 4: Seal in Humidity

Place your container in a humidity enclosure of your choice. I use an inexpensive plastic propagator box. Any clear box or clear plastic bag will do.

Step 5: Put the Snakes in a Pot

If you're propagating in water or any medium besides potting soil, once the roots are at least 1½ inch (4 cm) long, pot up the leaf blade sections in well-draining potting mix. Do this whether or not they've grown baby plants (pups) yet. Once they're in nutrient-rich potting mix, they will grow babies faster, regardless! If your leaves already grew babies, that's great too; pot them up with the original leaf still attached, as they're still obtaining nutrients from the leaf.

Step 6: Aftercare

Place your new plants in a spot with bright, indirect light, although they will thrive with a few hours of direct light per day and will begrudgingly put up with your lower-light shenanigans if that's what you have to give. Hey, we're all trying our best, right? Just know they won't grow much in lower light conditions. Let the soil dry out between watering and fertilize regularly if they are actively growing.

[above left] Three leaf blades with the water they grew roots in.

[above right] Snake plant pups! Closeup of results on one leaf.

SUCCULENT OFFSET PROPAGATION (WITH ROOTS)

Featuring *Haworthia fasciata* (Zebra Plant)

Just as with tropical plants such as calatheas, many species of succulents spread by producing offsets that shoot up from underground. Offsets are clones the mother plants asexually produce on their own. No Maury Povich "Who's the Daddy" episode needed! These offsets are free plants your mother plant created for you as a reward for keeping her happy. You can keep them growing in the pot, attached to the mother plant below the soil so your succulent grows into one big wacky clump, but because this is a propagation book, in this section we will talk about how to separate them so you can divide your plant into multiple healthy plants.

There are two different types of succulent offsets: those that have roots already attached and those that do not. When you remove an offset prior to it growing roots, propagate it similarly to how you would any beheaded succulent or succulent stem cutting that doesn't have roots (see instructions on page 142). If you wait for the pups to have roots before separating them, which is recommended when possible, making new plants will be a matter of division and your chances of success will be much higher.

Here are instructions for removing offsets that *have their own roots* from a haworthia.

Timeline

- 15–20 minutes for division and repotting
- 3 weeks–2 months, depending on the environment for a pup to adjust to a new pot and start growing again after repotting

Ideal For

- Aloe vera, haworthia, snake plant, agave, kalanchoe, echeveria, hens and chicks (*Sempervivum*)

Supplies Needed

- Clean cutting shears
- Succulent potting mix
- Pots with drainage holes: pots should be roughly ½–1½ inch (1–4 cm) wider than the roots of your plants in diameter

[page right] Two haworthia mommies with babies attached.

Preparation

- Ideally wait to propagate until the pups are 1½–2 inches (4–5 cm) tall with their own roots to increase the chances that they will survive the transition.

- Division is best done in spring or summer when it is actively growing, but if you are using grow lights or your plant is growing year-round, you can do this at any time.

- Ensure that your plant is well hydrated prior to division. If it has not been watered in a while, water your plant a few days prior to division.

Instructions

Step 1: Remove the Plant
Gently pull the plant out of its original pot.

Step 2: Separate the Offsets
Wiggle and break away the pups you want to propagate from the mother plant. If they don't come free easily, use a knife or cutting shears. Try to break as few roots as possible. While you're at it, prune off any dead, brown, shriveled foliage. It's not doing anybody justice!

3

4

Step 3: Allow Pups to Callus

Set the pups aside in a cool area with indirect light and good airflow to callus over for 1 to 2 days. This helps prevent rot later on.

Step 4: Repot Party!

Pot all other pups into new pots with drainage holes. I like to cover the drainage holes first with a piece of window screen or a pot screen made for this purpose, so no potting mix spills out. Make sure your pot is big enough to fit all the roots, and that you can cover them completely with succulent mix. Fill the pot up with mix up until the stems, but **do not cover the stems**, or you risk rotting them. Leave some space between the top of the potting mix and the top of the pot for water to pool when you water it.

Step 5: Aftercare

Do not water right away. Wait a few days or more to let all the cuts callus over even more before watering. For the first time watering at least, I like to bottom water to ensure that the potting mix gets thoroughly saturated. Place the plants in a warm spot, in a window with mostly bright, indirect light. They can tolerate a few hours of direct sun a day, such as in an east- or west-facing window. Once established, only water haworthia when the potting mix is bone dry, which in most climates will be every 10 to 14 days.

Cactus pads, also known as paddles

CACTUS PAD PROPAGATION

**Featuring Bunny Ear Cactus (*Opuntia microdasys*)
and Other Opuntia Species**

Now onto a thorny topic . . . propagating cacti pads, also known as paddles. While these thick spikey pads may look like leaves, they actually are modified *stems* used for storing water, performing photosynthesis, and, as we will review here, reproduction. These paddles are found on cacti in the genus *Opuntia*, commonly known as the prickly pear, of which there are 150 to 180 different species.

Prickly pear cacti are not only beautiful, but they're used for their medicinal properties in Mexican cultures. Those native to Mexico and other parts of the Americas are also used commonly in cooking and known as nopales. They are nutritious and high in fiber and antioxidants!

I'm sorry to say, however, that I am a terrible chef, and therefore in this section we'll only discuss how to propagate cacti pads, not cook them. Since we begin our propagation of pads by removing them from the cacti and pads do not have their own roots, these instructions will help you to propagate any succulent offsets without roots.

Now, let's take a look at how to propagate the popular bunny ear cactus, *O. microdasys*. The most ideal time to propagate this cactus is in spring, and the least ideal time to propagate would be during the winter when the plant is dormant, because it will be most susceptible to rot. However, I always say to give it a try any time of year indoors. Simply do your best to emulate the hot, dry conditions that this cactus thrives in wherever in your home that you propagate it, using grow lights, heating pads, etc. Remember: if you rot your cactus pad, don't feel bad. Get right back on that horse.

Timeline

- A few weeks to a few months depending on the species.

Ideal For

- Prickly pear cacti (*Opuntia* spp.)

Supplies Needed

- Leather gloves
- Silicone tipped or BBQ tongs, thick paper bag or newspaper
- Cactus/succulent potting mix
- Pot with drainage hole preferable

Preparation

Don't be a *prick* and put on thick leather or leatherlike gloves, use BBQ tongs, and/or use thick paper to grab the plant. This bunny cutie may not look threatening since it does not have huge spines like other cacti. However, those tiny white or yellow clusters of barbed bristles called **glochids** throughout its surface are no joke and are definitely worse than large pokey spines. These will stick all over your fingers and clothing and make you wish you never worked your way past the tropical plant section of this book.

Before you wish them away, however, know that your little bunny cacti use these glochids to deter animals in the wild and collect moisture from mist in the air, since there is so little rain in the desert. What a badass adaptation for something that looks like a tiny fuzzy animal!

Tip

If you get glochids on your fingers, use masking or packaging tape to get them out. If you get too many on your clothes, you may have to throw your clothing item away.

Instructions

Step 1: Slice the Pads Off

Either break off a pad using your tongs or slice off the pads with a sterilized knife or blade, or cut with scissors.

Step 2: Let the Wounds Callus

Lay the pads in a safe place (away from curious kids and pets) to callus over for 1 to 3 days to prevent rot.

Step 3: Pot Up the Pads

Use a cacti or succulent mix. You want to start off with the mix being dry (do not water it yet). Choose a pot with a drainage hole deep enough so that your pad can stand upward when you plant it and not fall over (usually about ⅓ of the pad buried). Or, if you find that your pads keep tipping over after planting, simply lay them down flat on top of the potting mix. Either way, roots can still grow from the cactus.

This is not a dish you want to serve to anyone, except your worst enemy.

3

4

5

Step 4: Environment and Watering

Place your pad propagation in a warm area that gets bright, indirect light. It is not recommended to put it in direct sun yet. How much and how often to water during cactus pad propagation is up for debate in the cacti world and people are successful watering different ways, depending on their growing conditions. This part is truly going to be trial and error for you. If your cactus pad rots, know YOU ARE NOT ALONE. We learn the most from our plant failures, so pick yourself back up and pick out another prickly pad and try again.

Here are three different options for watering, adapted from advice received from seasoned cactus growers:

1. **Sir-Mist-a-Lot**: I truly only recommend trying this method if you are in a warmer part of the country and it's spring or summer. Give the top of the soil a good misting to start off (not a deep watering). Avoid getting the pad itself wet. Continue to mist every time you see the soil dry out until the pad is fully matured (which is months beyond when it starts to root). If your cactus pad rots this way, then you know you used too much water and this method may not be best for your environment.

2. **The Middle Path** (what I do): "Tease" your pad with water. Mist the very edges of the pot just slightly every once in a while, when the soil is completely dry. The idea behind this is that you want the cacti to "sense" water, which in turn will trigger the plants to grow roots to reach the water. You don't want the cacti pad to physically feel the water, however.

3. **Total Drought:** A cactus pad can't rot unless there's too much moisture or humidity! Therefore, some growers don't water or mist at all until after roots form. They advise to keep the plant in a shady location while waiting for roots to form. After they do, move the plant to a brighter location and begin watering or misting sparingly.

Step 5: Wait for Roots

Shown here is another cactus of mine, a long spine purple prickly pear (*Opuntia macrocentra*) starting to root.

Roots will more often emerge from the **areoles** of the cactus than the base, and no, I'm not speaking about nipples here. Areoles are the small cushion-like spots all over the cactus from which the spines, glochids, and new growth emerge. A large pad will generally take longer to root than a small one, but the exact amount of time it takes depends on the species, how often you water, and your environmental conditions.

After a few weeks, you can test to see if roots have grown by wiggling and (gently) pulling your pad upward. If it resists, the root ball has likely established. Or you could be completely impatient like me and take the pad out to check for roots every week. This is not ideal if you can resist doing this, but I grew up on the East Coast and I'm not good at delayed gratification.

Step 6: Aftercare: Try Not to Rot your Cactus
Once your pad has roots, move it to a spot that receives direct sunlight at least a few hours a day (a south- or west-facing window). Water sparingly only when the soil is bone dry—approximately once every 2 to 3 weeks during the spring and summer and approximately once a month during the fall and winter. Fertilize with cactus and succulent fertilizer when it's actively growing (typically spring and summer or year-round if under grow lights).

Be patient as you wait for new pads to form. Often, they won't form until a year after rooting.

Key Points

- A dry cactus will not rot.
- A cactus can absorb moisture from the air (keep in mind especially if trying to grow in a more humid environment than the desert).
- Cacti store water that they need in their pads.
- An underwatered cactus can prune and shrivel or turn brown and dry and callus. If this happens, give your cacti a thorough watering.
- Cacti grow in direct sun in the desert.

Scion

Rootstock

GRAFTING

One early predictor of a plant propagation lifestyle (yes, it's a whole lifestyle) is that, as a kid, you had a fascination with the little candy-colored cactus orbs fused atop slender green cacti bottoms at the plant nursery. Bright and cheery, they looked (and still look, as they are still being sold and have stood the test of time) like tiny microphones you would see in a Trolls movie. These succulents are not one plant, but two that are individually cut then secured together so they unite and grow as one. This unique type of asexual propagation is called **grafting**. Like a good marriage, when grafting a succulent, a union is formed between the two plants and they continue to grow as one individual, sharing resources. (They would make excellent wedding favors, don't you think? A true symbol of union! Don't forget to name your firstborn after me if you use this idea.)

Grafted succulents do not thrive on marital bliss, however. Rather, their survival depends on the bottom plant species (the rootstock) sharing the water and nutrients it acquires with the top species (the scion). I suppose my own marriage wouldn't survive if my husband didn't share his snacks with me. While I do not cover the steps on how to graft a succulent in this book, as it is not my forte, I do encourage anyone who is interested to give it a try. Note, however, that for grafting to be successful, the two plants have to be closely related, genetically. You can drop the marriage analogy at this point.

TRENDING TECHNIQUES

Love it or hate it, social media effects the way we dress, the music we listen to, the things we buy, and even the way we propagate plants. (If you have not yet ventured to the side of the internet that relishes a good houseplant hack, I highly recommend giving it a try.) The online houseplant community will never tire of new plant care information, especially if it enables them to take care of, or reproduce, their collection more efficiently.

Although plant propagation videos will never go as viral as a Britney Spears Instagram post or break the internet like a Beyoncé album drop, the techniques discussed in this section were all the rage when they were first posted on social media. Not only did each have their 15 minutes of internet fame when first introduced to the world, but their utility has given them real staying power.

Each of these three methods are popular for a reason—they enable you to create more plants than you ever thought you could, from one mother plant. And, if there is one thing we all know about plant people, it's that deep down, even if we know we don't have any more space, time, or money, we all want more plants.

STRING OF HEARTS BUTTERFLY METHOD

In 2020, a houseplant content creator posted a video to Instagram demonstrating a unique way to propagate string of hearts, and the plant community was never the same. In her video, she called this technique the butterfly method because the leaves splay out when you flatten them down on to the soil, similar to when you butterfly a chicken breast. I'm not a big fan of thinking about raw chicken, but at least it has a better ring to it than if she had called it the spatchcock method. Technically, it's known as a two-leaf bud cutting which is a type of stem cutting.

This method can be used on all cultivars of string of hearts, including the variegated string of hearts, 'Orange River' and, as seen in this section, string of hearts 'Silver Glory'. The reason I love this method is because no node is left behind. By propagating every single node, you maximize the number of new plants you can make from a single vine.

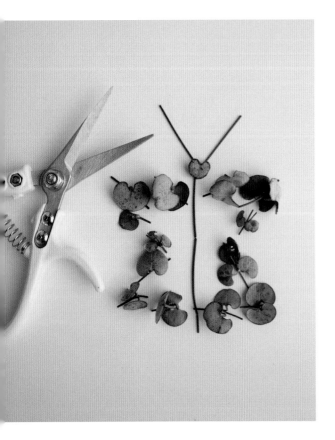

This method has nothing to do with actual butterflies. It is named as such because the technique mimics the way you butterfly chicken for cooking.

Supplies Needed

- Sterilized cutting tool
- Rooting hormone gel or powder (optional)
- Well-draining succulent and cacti potting mix
- 2–4" (5–10 cm) pot
- Plastic zipper-top bag

Timeline

- 1.5–2 months to start of new growth
- 3–4 months for established new plant

Ideal For

- All string of hearts cultivars

Preparation

- Propagate your plant within a week of the last watering to ensure that your plant is fully hydrated.

Instructions

Step 1: Choose a Vine

Select healthy vine(s) with leaves that feel firm and healthy, not flimsy and underwatered, or mushy and rotting.

Step 2: Isolate the Nodes

With clean, sterilized cutting shears, cut the vine directly before and after each node. You don't need any of the vine for the propagation to be successful, only the leaves and the point at which they meet: the node. It's okay if a tiny bit of stem remains. It will often decompose during propagation. You may notice some nodes already have an aerial tuber attached. That is great. Nodes with aerial tubers typically grow roots and shoots faster than their bare-naked node counterparts.

Step 3: Set Up Your Container

Some people like to propagate string of hearts nodes in moist sphagnum moss in a plastic takeout container with a lid and then transfer the rooted vines to a pot with succulent mix after they've rooted. Using sphagnum moss minimizes the chance of rot if your environmental conditions aren't just right. However, I have found great success propagating them directly in a moist, well-aerated potting mix, exactly as shown. When propagated this way, you won't have to transfer them and stress them out, nor will you have to deal with picking moss off tiny delicate roots. There are benefits to both ways!

I like to use the pot I intend on growing them in for the propagation. This way, there is no need to transplant at the end. In the photos, I am using ⅔ store-bought cactus and succulent mix straight out of the bag with ⅓ perlite added in. I made my own pot to grow them in by poking a drainage hole into the bottom of a plastic cup. The transparent cup enables me to watch the roots grow. If using a nursery pot, I recommend starting with a 3 to 4 inch (8 to 10 cm) pot.

Water the mix so the top third of the mix is saturated. I find it best **not** to soak the potting mix all of the way through as these are succulents without roots, and remember: you want to "tease" them and make them work for their water. Send the message to your succulents: Grow roots to get that bit of water that's available, or you won't survive.

Tough love, baby!

Step 4: Plant the Nodes

It's optional to first dip each node into rooting hormone gel or powder. Then, gently press each node down onto your substrate, with the top of the hearts facing up, purple sides down. This is where the "butterfly" comes in: each will resemble a butterflied chicken breast. Each node should be pressed into the potting mix or at the very least grazing it.

Step 5: Seal in the Humidity

Cover your container to lock in as much humidity as possible. You can use any type of humidity chamber, such as a propagator, a homemade propagation box, or a simple sealed zipper-top bag as shown in the photos. Feel free to open your humidity chamber to let air in occasionally, but don't forget to close it back up soon after. I find that letting air in is not often necessary unless mold starts to grow. When I do open the bag or box, it frequently ends with me forgetting to close it back up again and losing my propagation project. (Anyone else with attention deficit issues?) If you do notice the soil beginning to dry out at any point in time, remoisten it. After 1.5 to 2 months, you will see roots and new baby leaves begin to form.

Step 6: All Done!

After 3 to 4 months, you can remove the plant from its enclosure. Ideally, all of the hearts are firm and the nodes have white roots growing into the substrate.

Step 7: Aftercare

Place your new string of hearts in a sunny window. They can tolerate a few hours of direct sunlight per day. Allow the substrate to dry out 100 percent before watering or the hearts will rot. When you water, do so very thoroughly. Fertilize at least through spring and summer, but I fertilize mine year-round because I am an overachieving heart grower.

ZZ PLANT LEAFLET PROPAGATION

Maybe it's the magical pearl-like rhizomes the leaves grow at their base, or maybe it's the satisfying way you can line the shiny, ovate leaflets up in soil like potato chips sticking out of a bowl of dip. Perhaps it's the way each leaflet resembles an old-fashioned quill pen dipped in a glass jar of ink when sitting in a small propagation vessel filled with water waiting to grow roots. No matter how you look at it, ZZ plant leaflet propagation is as beautiful as it is productive.

The most obvious way to propagate a ZZ plant is to simply pull apart the long stalks and make new plants through division of the rhizomes. However, much like the string of hearts butterfly technique, if you're looking to maximize the number of new plants from your existing ZZ plant, leaflet propagation is the way to go. This method gained popularity a few years ago when someone on social media demonstrated that ZZ plant leaves, when submerged in water for an extended period of time, grew rhizomes and tiny roots on the bottom. Instead of making just a few new plants from your ZZ plant, you could make 12 to 20 from one petiole. This marked the start of a new houseplant propagation craze.

The one thing you should note, however, is that this method does take 7 to 9 months before the leaves are ready for planting with new rhizomes and roots. Therefore, if having patience is not your strong suit, you may want to try a different technique (such as division, or the credit card propagation technique where all you have to do is swipe it at the plant shop and you get an entirely new plant in seconds).

Before we get started, here is the anatomy of a ZZ plant above the soil level:

[above] This unique propagation technique is shown here on the black-leaved ZZ plant 'Raven'.

[page right] ZZ plant leaves are called *leaflets*. ZZ plants do not have traditional stems. They grow directly out of a round tuberous rhizome under the soil, and the main stalk is called a *petiole* below where the leaves grow and is referred to as a *rachis* (RA-kis) along where the leaflets grow. The itty-bitty stalks attaching the leaflets to the rachis are called the *petiolules* (PET-ee-ole-lules). I know, it's not confusing at all!

Rachis (spans the area that contains the leaflets)

Leaflets

Petiolules

Petiole (the stem below where the leaves grow)

1

2

3

Timeline

- 2–3 months for initial rhizome and root development
- 7–9 months for new rhizomes to be ready for transplanting

Ideal For

- All varieties and cultivars of ZZ plant (*Zamioculcas zamiifolia*)

Supplies Needed

- Sterilized cutting instrument
- Rooting hormone powder
- The patience of a saint
- New pots for the babies

For Water Propagation:

- Glass vessels for water propagation
- Fresh succulent potting soil will be needed in 7–9 months

For Soil Propagation:

- Succulent potting soil
- A container or pot for the soil

You can propagate the leaves in a variety of substrates, but in this section, we will review water and soil.

Instructions

Step 1: Cut Off the Leaflets

Remove the leaflets you would like to use by cutting them at the base of their petiolule.

Step 2: Rooting Hormone

Dip the petiolule into rooting hormone gel or powder. This step isn't necessary, but believe me, this type of propagation takes so long that you'll be grateful for anything that helps speed it up.

Step 3: Insert in the Growing Medium

Place the leaflets in the moistened potting medium of choice. As always, water is a safe bet and easily accessible, and straight-to-soil propagation in a humidity enclosure such as a sealed plastic bag or in a prop box will often (but not always) produce more robust results, faster.

Step 4: Wait for New Plants to Grow

If growing in water, change the water at least every few weeks and wash your glasses out to clean out any algae.

Rooted in water

Rooted in moist potting mix and perlite

4A

4B

5

[top] Three Month Results: You can expect to see small rhizomes and little roots sticking out of the leaflets to various degrees depending on the vigor of the leaflets and your environmental conditions.

[bottom left] Five Month Results: If the roots aren't at least 1 inch (3 cm) long and there is no shoot starting to come up yet, keep waiting! At this point you may be convinced you're actually just growing one of those marimo moss balls attached to a leaf, but a ZZ plant shoot, or multiple shoots, will eventually sprout up. It's been so long that these leaves are now officially pets. I hope you named them.

[bottom right] Nine Month Results: Move on to Step 5!

Step 5: Pot Up the Rhizomes!

Once the roots are at least 1 to 2 inches (3 to 5 cm) long and the shoots have begun to develop, pot up your rhizomes in a well-draining potting soil. A succulent mix straight out of the bag should work well with about 20 percent extra perlite added in for more drainage. Bonus if you mix some super gritty succulent mix as well. Begin to water as you would a mature ZZ plant, allowing the soil to dry out between watering. Give lots of bright, indirect light.

Can you see why this is named *Alocasia baginda* 'Dragon Scale'?

ALOCASIA CORM PROPAGATION

Back around the year 2020, a a friend and fellow collector by the name of Sarah Spaulding (@botanophile on Instagram) struck plant propagation gold: she uncovered the article "Propagation of Alocasias" written by Jim Georgusis published in the International Aroid Society's newsletter *Aroideana* in the 1980s from the society's archives. Prior to this, how to propagate an alocasia other than through division was not widespread knowledge among the current generation of alocoasia collectors (I'm sure some people knew, but it wasn't common knowledge). In this article, Georgusis describes how to propagate an alocasia using the bulb-like structures, or corms*, that the plant produces beneath the soil. Sarah shared this article with prominent collectors, growers, and influencers on social media, and the technique went viral.

*While these structures are more technicially aligned with the descriptor tubercules *as described in chapter 1, we will continue to use the word corm here as that is what is most widely used in the houseplant hobby and used by Georgusis in his article.*

As a result of this technique going viral, some entrepreneurs and influencers created online shops dedicated to selling alocasia corms and the alocasias they grew from them. This was a profitable business for a couple years as the plants were in high demand and the corms were easy to harvest. However, after the pandemic tissue-culture laboratory boom, which rapidly flooded the market with supply and drove prices of fully grown alocasias down below $10 a piece in many cases, these small businesses shifted to selling other plants. In the time it takes to grow a few alocasia by corm, you can grow hundreds by tissue culture. Today, most of the alocasia you see for sale are tissue culture grown by the hundreds or thousands in laboratories and sold in large plant shops and big box stores. Alocasia propagation by corm still remains an amazing way to multiply your collection, share it with friends, and even sell or trade rare specimens. I should also note, however, that selling rare, variegated alocasia corms can still be very profitable.

As a review, corms are the survival structures of the alocasia. If the main mother plant dies, it's the plants that sprout from the corms that ensure survival of the plant as a whole. However, when you have a young healthy alocasia in a pot with many corms still attached to the mother plant, the corms are still recieving nutrients from the mother plant. They don't recieve any signal that there's an ugent need to grow roots and sprout. Therefore, as any grower will attest, some corms will eventually sprout on their own, but it takes quite a while and how long depends on your growing conditions. The more ideal your growing conditions, the more likely new plantlets will sprout from these corms. Otherwise, in most ambient household growing conditions, many of the corms simply remain dormant, nestled like little eggs within the mother plant's roots.

This is why I think of corms like (some) adult children that need tough love. Push them out of the house, tell them to get a job, and to start paying bills if you want them to be productive sooner. When you separate a corm from the mother plant, the storage structure will then grow roots and shoots to survive. It's sink or swim out there in the big world, little corm! It's amazing how fast new growth will happen after separation, given the right conditions.

So, that is where our propagation journey begins: finding these corms and telling them to go out on their own and get a job growing roots and shoots!

Timeline

- 6–9 weeks for the first leaf to begin to sprout
- Roughly 8–12 weeks until the first leaf unfurls and there is sufficient root growth for transplanting. Timing varies depending on individual corms, alocasia species, and environmental conditions.

Ideal For

- Alocasias

Supplies Needed

- Clear container with a lid (i.e., takeout container or seedling tray)
- Substrate: sphagnum moss, Fluval Stratum

[top] When you go corm spelunking, you may find unsprouted corms, or corms that already have roots and/or shoots.

[bottom] An alocasia corm sprouting roots.

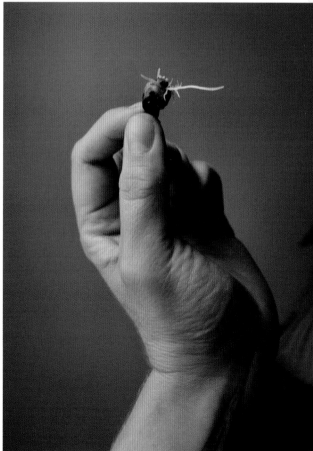

1A

1B

2

Instructions

Step 1: Corm Treasure Hunt

The best way to find corms is to first gently remove your plant from its pot. Remove some of the potting media with your fingers so you can get a better view of the roots and potential corms. Feel and look around the upper third of the root mass for small round corms (that's where the majority will be hiding). Use your fingertips to search both the outer edge of the root ball and inside the middle of the root mass. If something jumps out and bites your finger: RUN!

Kidding.

Tip

If you break a root or two, that's okay. The plant will grow new ones. If it makes you feel more confident to handle your plant's roots, know that often when I repot a healthy, fully mature alocasia, I purposefully cut off at least the bottom ¼ of the root system simply because they grow excessive roots and I need to be able to fit them in reasonably sized pots. In fact, in Georgusis's article, he instructs the reader to reduce the root mass of the mother plant by half before replanting again!

Also, if you see any roots looking sus while you're down there, chop them off. Rotting beige, black, brown, and mushy roots must be cut away.

Step 2: Corm Extraction

When you find a corm, simply snap or twist it off its stolon with your fingers. It does not matter whether you keep any of the brown stolon attached to the corm. I typically break it off.

Tip

While most corms are brown and round with a pointed tip like a teardrop, the exact size, shape, and color of the corm depends on the species. For instance, the corms of *Alocasia cuprea* are pink inside. Also, some species of alocasias grow corms directly off the main rhizome/ underground stem without a stolon, as is the case with *Alocasia* 'Kapit'.

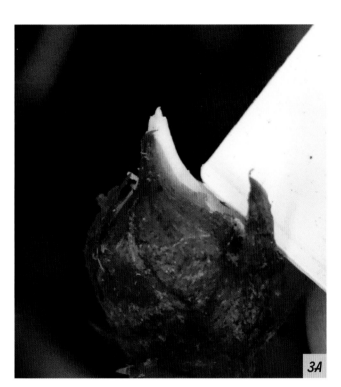

3A

New roots will grow from the top of the corm!

3B

Step 3: Peel the Corm

This step is optional, but it speeds up the rooting process. Remove the brown outer scales the roots need to poke through to grow. You can use your fingernail, but it's even easier to use the edge of something thin and plastic, such as a credit card or garden tag. Take care not to damage the corm's interior tissue. What's left is the tissue that surrounds the **apical bud**. On the corm, the apical bud is the pointy tip from where the shoot grows.

New roots will form toward the top of the corm on this tissue in a ring around the apical bud. Strange, I know. I bet you expected roots to grow out of the bottom of the corm!

Step 4: Prepare Your Propagation Container

Prepare to plant the corm in the medium of your choice in an upcycled takeout container, seedling tray, or even plastic bag. My personal preference is to use chopped up sphagnum moss mixed 50/50 with perlite, or Fluval Stratum mixed 50/50 with perlite. In both cases, the substrate must remain moist throughout the entire propagation period. How moist? It should be as moist as a gently wrung-out sponge. If you see it start to dry out, remoisten it! (Look, it's not my favorite word either, but it works).

Step 5: Lay in Substrate

Plant your little nuggets, leaving only the very tip of the corm exposed.

Step 6: Cover, Place Under a Light Source, and Wait for Growth

Seal the top of the container to maximize humidity and put the corm(s) in a window with mostly bright, indirect light or under a grow light. I'm placing mine on a shelf, 1 foot (30 cm) under two 3-foot (91-cm) long T8 LED grow lights.

Step 7: Observe Growth

If you have a healthy, peeled corm that receives the proper warmth, light, humidity, and moisture, in 2 to 4 weeks your corm should have roots similar to this. In 6 to 9 weeks, leaves will begin to sprout.

Step 8: Pot Them Up

After the first leaf has fully unfurled, and the roots are at least 1½ inch (4 cm) long, your corm is ready to be transferred to a potting mix or growing medium of your choice. In my personal growing experience, alocasias do well in an aroid mix in a traditional pot, and even better in pon in a self-watering pot.

If transferring to a different substrate than the one you propagated in, tease away the propagation substrate from the roots and pot in the new media. Like any delicate plant going through a substrate transition period, after potting your new alocasias (especially if working with a sensitive jewel alocasia species), it's a good idea to keep them at around 70 to 80 percent humidity for a couple weeks while they adjust to their new medium before placing them where you intend on growing them. Or, just keep them in 70 to 80 percent humidity for life and your alocasia will love you forever.

Step 9: Aftercare

Give plenty of bright, indirect light. If they are stretching toward the light, that means they're not getting enough. Humidity above 50 percent is ideal, but hardier species can tolerate lower humidity. When you water, water thoroughly and remember that alocasias don't enjoy drying out completely between watering. They are heavy feeders and if actively growing, fertilize year-round.

6B

7

8

Chapter 5

PROPAGATION AS REHABILITATION

Why is this chapter going to be as good as therapy?

We are going to get to the root of all your problems.

Propagation isn't just for making free baby plants. It's also used to fix and maintain the ones you have. Just as with humans, plant surgery can be medical or cosmetic in nature. We "chop and prop" to help plants that are sick and dying and those that need a complete aesthetic refresh. (No plant Botox though).

Now that you have mastered houseplant anatomy and propagation techniques, you can apply that information to give your mature plants the physical rejuvenation they are yearning for, or to save a houseplant in your botanical ICU.

COSMETIC PLANT PROPAGATION

There are untidy plants you can clean up with a little pruning. Then there are the hot-mess-express plants you need to hit the total reset button with. Mature, climbing philodendrons are no stranger to this second category as it is common for their bottom leaves to turn yellow and fall off over time, leaving behind a long, bare stem. To refresh philodendrons that look long and lanky or top-heavy, it is common practice to chop up the plant into pieces and regrow them into individual plants. The bottom, rooted stump will regrow all new foliage!

For example, most recently I chopped up my *Philodendron* 'White Wizard' into individual nodes because as it aged, it started to look droopy. I felt its pain. I put each node in a propagation box to root on top of moist sphagnum moss. To learn how to propagate a plant by node cutting with more in-depth step-by-step instructions, see the Tropical Plants Node Propagation section on page 84.

PROPAGATION AS A LIFE-SAVING MEASURE

Another critical application for plant propagation is to rescue sick plants or those teetering on the brink of death. This is common for plants such as alocasia, which have a reputation for, as one Botanist puts it, "sudden collapse." When you start to see the leaves of your well-watered plants drooping, wrinkling, and/or turning funky colors and maybe even falling off, you have three choices:

1. Gift the plant to your enemy.

2. Compost it.

3. Try to save its life.

I'm sorry to disappoint you, but we aren't going to review number one. We're going to skip to number three, make like The Fray in the year 2006 and practice "How to Save a Life." Lucky for you (but unlucky for me), I have a hoya that was not looking so hot to use as an example. Read on to follow step by step if and how I can save my struggling *Hoya nicholsoniae* 'New Guinea Ghost'.

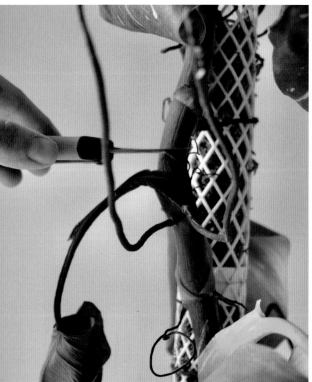

I'm giving my *Philodendron erubescens* 'White Wizard' a total makeover. I cut the plant down into individual nodes and will root them in a propagation box.

Re-aliving my *Alocasia infernalis* 'Kapit'.

Hoya Root-Rot Rehab
Featuring *Hoya nicholsoniae* 'New Guinea Ghost'

My once-magnificent *Hoya nicholsoniae* 'New Guinea Ghost' is trying to become a literal plant ghost. However, I'm not about to let that happen. First, when any plant is struggling, assess whether you can stop the struggle bus the plant is riding, or if it is on a one-way street to the compost. Like any good doctor, first gather your tools, perform a full assessment, diagnose, and then treat.

Supplies Needed

- A magnifying glass
- A scalpel or cutting shears
- Rubbing alcohol
- Rooting hormone powder
- Fluval Stratum and perlite (or LECA or Lechuza pon)

Preparation

- **Observation 1: Is the plant still green?**
  Assess whether there is healthy green plant tissue left to save. If your plant is completely brown and crispy, it's not coming back. Brown and yellow dying plant tissue does not turn green again!

- **Observation 2: Foliage pest inspection.**
  Grab your magnifying glass. Ideally you have a 45x magnifying glass with a light, a jeweler's loupe, or a handheld microscope, but use whatever you have. Through the highest magnification, with the light on (if it has one), search the top and undersides of the leaves and stems for pests. Look for webbing, larvae, shed exoskeletons, and adult pests. If any are found, identify and treat appropriately. Cut off any leaves that show signs of severe infestation at the base of the leaf.

- **Observation 3: Foliage health checkup.** What are your plant's leaves trying to tell you?

  - **Yellow/Brown Leaves**
    If the **oldest leaves** of your plant are turning yellow, and you don't see pests with a magnifying glass, the yellowing is most likely natural, due to a process called leaf **senescence**. This is when old leaves die and the plant reabsorbs nutrients from those leaves. You can let those yellow leaves be.

    However, if the **newer leaves** emerge looking like a Salvador Dalí painting, they're browning, have spots, or contain colors that represent a tie-dye T-shirt (and your plant is not a croton), you have cause for some concern. Evaluate your foliage for signs of over- or under- or inconsistent watering and nutrient issues. Take note of anything that looks like a bacterial or fungal outbreak or virus.* While these are less common than nutrient deficiencies and environmental stresses, they can occur.

    Pathogens will typically continue to spread throughout the plant over time. Therefore, if you truly think your plant has a fungal or bacterial outbreak, after ideally confirming with experts, make like Uncle Joey from the show *Full House* and CUT IT OUT! (I know, I just aged myself).

1

Thin, dried up roots

2

Instructions

Step 1: Serenade Your Plant

This is critical for the rehabilitation process. This is my Justin Bieber *Ghost of You* remix, which you can use if you cannot come up with a song of your own:

> So if I can't get close to you
> I'll settle for 'New Guinea Ghost' from you.
> I love plants more than life.
>
> And if you can't be next to me
> You can send all of your plants to me.
> I love plants more than life.

Step 2: Root Inspection

My hoya's leaves started to whither even though it was well-watered and had no signs of pests. Therefore, I knew there must be something wrong with the root system. I pulled it out of the pot, and sure enough, the roots were desiccated.

Likewise, you will want to take your struggling plant out of its pot to assess the root ball and confirm that the roots are white to cream in color, or the color they should be depending on the species. Some plants have pink or red roots. If they are black or very dark brown, however, they are most likely dead.

Next, check if the roots are firm with a healthy outer sheath. When you gently run your thumb and forefinger along a single root, it should remain intact. In alocasias and hoyas, for example, it is common for roots to remain cream in color even when rotted, but when you gently run your thumb and forefinger along a damaged root, the outer sheath will slide right off.

What are your plant's roots trying to tell you?

Sterilize the cutting shears after each cut with a rubbing alcohol to avoid spreading pathogens.

*Note: The only true way to diagnose a virus, bacterial, or fungal outbreak is by sending a leaf sample to a lab or ordering an at-home testing kit from a laboratory. However, right or wrong, many growers assume the presence of spotting typically associated with these infections is reason enough to start treating the plant with antifungals or antibacterial sprays (or if a virus is suspected, disposing of the plant). Others take a more conservative approach and work to control the environmental factors around the plant to see if that resolves the issue on its own.

- **Withered Leaves**
  If the leaves are wrinkled, but the plant is well watered, more than likely the roots are damaged and not able to absorb water. This is where propagation as a rehabilitation tool comes into play. Proceed to the following instructions.

ROOT CONDITION	VERDICT	ACTION
Firm and white	Healthy	Keep
Thin and black or brown	Rotted	Remove
Black or brown and mushy	Rotted	Remove
None, as in they already melted off	Rotted AF	Cry. Then remove that area of stem.
White or beige but the outer sheath slides off when you run your thumb and pointer finger along the root	Rotted	Remove
White or beige and shriveled	Desiccated/overly dried out	Remove
White spots/white powdery insects on the roots and inner pot	Root mealybugs	BURN EVERYTHING. Kidding. Remove or keep and treat.

This is what classic root rot on an alocasia looks like. It looks like I am holding dirt in my fingers, but that's a mass of dark brown, rotted roots.

3A 3B

Step 3: Remove the Rot

I ended up cutting off all the roots from every single vine in the pot. ALL. OF. THEM. The end results were a bunch of naked hoya cuttings. I then removed the bottom leaf from each stem so that there would be a few inches of healthy, bare stem to insert in the substrate. From the bare stem, new roots grow!

Your goal here is to cut off every part of the plant that is showing signs of rot. If there are only a select few roots that have rot, cut them all off with a sterilized knife or cutting shears. If any part of the above or underground stem is showing signs of rot, cut that part of the stem completely off. It's a good idea to sterilize your cutting shears with rubbing alcohol between cuts if making multiple cuts in areas that contain rot, to avoid spreading pathogens.

Step 4: Regrow Roots

Place your fresh cuttings in the substrate of your choice.
If you need help choosing, refer back to chapter 3. In the
accompanying image, I rehabbed my hoya in a mix of
Fluval Stratum and chunky perlite (mixed not-so-evenly
because to be honest I was just tossing in handfuls). These
two substrates are popular for hoya propagation. Plants
grow roots quickly in Stratum due its high nutrient content
and ideal pH, and the roots that form are acclimated
to a solid substrate (as opposed to roots acclimated to
water alone).

As shown in the image for step 4, this is a temporary
semi-hydroponic setup. The glass vase does not have a
drainage hole. I simply keep a couple of inches of water
in the bottom of the vase at all times and refill it when I
notice the water level getting low. The water at the bottom
is not touching the stems of the plant (but eventually
the roots may grow into it). This type of setup is just fine
for a temporary propagation situation, but I would not
recommend it for long-term semi-hydroponic growing
where you need to flush the water of excess salts and check
pH levels.

If you can place your setup in a prop box, cabinet,
humidity chamber, or near a humidifier during this time,
that would be beneficial to increase humidity and speed
up results.

Step 5: Repot and Start Fresh!

Shown here is hoya root growth in Fluval Stratum and
perlite after just one month. Do you believe me now
that Fluval Stratum is magic? After your new roots are
at least 1 inch (3 cm) long, you can pot your plant in
fresh substrate of your choice and resume your regular
care routine.

Now you know how to make new plant babies from
cuttings and other propagules, as well as through
pollination. You also know how to give botanical
makeovers and rescue some of the struggling plants
you may have. I hope this book enables and encourages
you to share more plants with friends and family. And
who knows, maybe you will be inspired to start a little
side hustle one day. Do whatever brings you—and helps
you spread—joy . . . and plants!

4

5

FURTHER READING

CHAPTER 1: ANATOMY FOR THE HOUSEPLANT PROPAGATOR

Capon, B. 2022. *Botany for Gardeners*. Portland, OR: Timber Press.

Boyce, P. C. 2014. "A Review of Alocasia (Araceae: Colocasieae) for Thailand Including a Novel Species and New Species Records from South-West Thailand." *Thai Forest Bulletin* (Botany) 36: 1–17. Retrieved from https://li01.tci-thaijo.org/index.php/ThaiForestBulletin/article/view/24171.

Hay, A. 2000. "381. Alocasia Nebula: *Araceae*." *Curtis's Botanical Magazine* 17, no. 1: 14–18. http://www.jstor.org/stable/45065411

Martin, F. W., & E. Cabanillas. 1976. "Leren (*Calathea Allouia*), a Little Known Tuberous Root Crop of the Caribbean." *Economic Botany* 30: 249–56.

Tomlinson, P. 2008. "Morphological and Anatomical Characteristics of Marantaceae." *Journal of the Linnean Society of London* (Botany) 58: 55–78. https://doi.org/10.1111/j.1095-8339.1961. tb01080.x

CHAPTER 2: THE BIRDS AND THE BEES OF HOUSEPLANT REPRODUCTION

Albre, Jérôme, Angelique Quilichini, and Marc Gibernau. 2003. "Pollination Ecology of Arum Italicum (Araceae)." *Botanical Journal of the Linnean Society* 141: 205–214. https://doi.org/10.1046/j.1095-8339.2003.00139.x.

Fernandes Cardoso, João Custódio, Matheus Lacerda Viana, Raphael Matias, Marco Furtado, Ana Caetano, Hélder Consolaro, and Vinícius L. G. Brito, Vinícius. 2018. "Towards a Unified Terminology for Angiosperm Reproductive System." *Acta Botanica Brasilica* 32: 329–48. https://doi.org/10.1590/0102-33062018abb0124.

Gibernau, M. 2003. "Pollinators and Visitors of Aroid Inflorescences." *Aroideana* 26: 66–83.

Marshall, T., M. Davis, A. McCaskill, and F. Corotto. 2014. "Spadix Function in Pinellia pedatisecta (Araceae)." *Aroideana Journal of the International Aroid Society, Inc.* 37, no. 1: 89–94.

McCormick, S. 2013. "Pollen." *Current Biology* 23, no. 22: R988–90.

Takano, K. T., R. Repin, M. B. Mohamed, and M. J. Toda. (2012). "Pollination Mutualism between Alocasia macrorrhizos (Araceae) and Two Taxonomically undescribed Colocasiomyia Species (Diptera: Drosophilidae) in Sabah, Borneo." *Plant Biology* 14, no. 4: 555–564. https://doi.org/10.1111/j.1438-8677.2011.00541.x.

Walker, T. 2020. *Pollination: The Enduring Relationship between Plant and Pollinator.* Princeton, NJ: Princeton University Press.

Zona, S. 2023. *A Gardener's Guide to Botany: The Biology behind the Plants You Love, How They Grow, and What They Need.* Boston: Cool Springs Press.

CHAPTER 3: TOOLS, MATERIALS, AND ENVIRONMENT

Lopez, R., and E. Runkle. 2005. "Managing Light during Propagation." *GPN Magazine*, December 2005. Retrieved from https://www.purdue.edu/hla/sites/cea/wp-content/uploads/sites/15/2005/06/17-Managing-light-during-propagation.pdf.

Pokorny, K. 2022. "Harvesting Peat Moss Contributes to Climate Change, Oregon State Scientist Says." *Oregon State Today*, December 9, 2022. Retrieved from https://today.oregonstate.edu/news/harvesting-peat-moss-contributes-climate-change-oregon-state-scientist-says.

Vanneste, S., and J. Friml. 2009. "Auxin: A Trigger for Change in Plant Development." *Cell* 136: 1005–1016. https://doi.org/10.1016/j.cell.2009.03.001.

CHAPTER 4: PROPAGATION METHODS

Croat, T. B. n.d. "The Sectional Groupings of Anthurium (Araceae)." Retrieved from https://www.aroid.org/genera/anthurium/anthsections.php.

Boyce, P. C. and T.B. Croat. 2018. "The Uberlist of Araceae, Totals for Published and Estimated Number of Species in Aroid Genera." https://www.aroid.org/genera/120110uberlist.pdf. Accessed December 2018.

North Carolina State University Extension. North Carolina Extension Gardener Handbook. Raleigh, NC: North Carolina State University. Retrieved from https://content.ces.ncsu.edu/extension-gardener-handbook.

Robert, B. (2009, August 10). "Propagating Opuntia." The Scott Arboretum of Swarthmore College. Retrieved from https://scott-dev.domains.swarthmore.edu/propagating-optunia/.

www.aroid.org

www.begonias.org

PLANT PROPAGATION EMERGENCY EXIT PAGE

You have either landed on this page in error or because you have decided that plant propagation is not for you at this time. While the grass and plant cuttings may not always be greener on the other side, perhaps you'd be more suited for knitting, skydiving, Pokémon cards, rock climbing, origami, jiu jitsu, or sourdough bread making. Perhaps ice baths and marathons are more your speed?

    Whatever you choose to do next with your time, know that plants will be here, waiting for you, literally rooting for you! And so will this book, (which by the way, makes an excellent Pokémon card and origami display platform). I have a feeling you'll be back one day.

RESOURCES

The following are products I use to propagate my own plants and recommend:

Potting Soil and Mixes

- FoxFarm Ocean Forest Potting Soil
- Very Plant's Molly's Aroid Mix
- Black Gold Cactus & Succulent Mix
- Bonsai Jack's Gritty Succulent Mix
- Very Plant's Molly's Succulent Mix

Fertilizers, Nutrients, Growth Stimulants

- SUPERthrive Foliage-Pro Liquid Plant Food
- FoxFarm: Grow Big Fertilizer
- TPS Plant Foods Succulent Plant Food
- TPS Plant Foods TPS-One, One Part Liquid Nutrient
- SUPERthrive The Original Vitamin Solution
- Real Growers Recharge - Natural Plant Growth Stimulant

Rooting Hormone Compound

- Clonex Rooting Gel
- Bonide Bontone II Rooting Powder
- Garden Safe TakeRoot Rooting Hormone

Grow Lights

- Soltech Solutions full spectrum LED lights
  (use code PLANTBABIES for 15 percent off)
- Sansi full spectrum LED lights
- Barrina T5 full spectrum grow lights

Potting Supplies

- RT1home Recycled Rubber Potting Tarp: rt1home.com/
- RT1home Metal Soil Scoop: rt1home.com/

ABOUT THE AUTHOR

Lindsay Sisti, better known as @alltheplantbabies across social media, lives in Oak Park, Illinois, with her awesome husband, Chris, and their exclusive hybrids: Sienna, Jasper, and Willow. These three specimens display hybrid vigor and great beauty. They are the greatest source of joy in her life.

Now that she is finished making little people, Lindsay can be found creating new tropical plant hybrids in her plant nursery. In 2023 she received a U.S. patent for the creation of *Alocasia* 'Green Unicorn', a cross between *Alocasia azlanii* and *Alocasia baginda* 'Dragon Scale'. She also successfully created a hybrid between *Alocasia azlanii* and *Alocasia chaii* and over a dozen anthurium hybrids (some of which are being bred into the F2 and F3 generations). She runs an online plant shop, All the Plant Babies, where she sells her unique specimens and propagations.

The daughter of a microbiologist and doctor, Lindsay was fully immersed in the world of science since childhood. Today, she blends her passion for entrepreneurship, education, and science together in her career as a plant shop owner, content creator, and hybridizer. While she loves to share knowledge with others, Lindsay is most motivated by the fact that she still knows so little compared to all the information yet to be taken in, and things yet to be discovered. Last but not least, she is a forever student of this incredible, magical, world.

Lindsay holding her fourth child,
Alocasia 'Green Unicorn'

ACKNOWLEDGMENTS

First and foremost, I'd like to thank my incredible husband, Chris. I'll never be able to thank you enough for doing all those bedtimes and weekends solo so that this book could become reality. You are the grounded moss pole to my climbing ambitions.

Thank you, mom. Having a feminist, scientist mother was one of the greatest sources of inspiration in my life. Dad, roaming through your plants in our home jungle growing up is one of my core childhood memories. Your unwavering devotion to every one of your plants (even the scraggly, dying ones that Mom hated) taught me the value of plant life from a young age.

Thank you to my childhood best friends Ali Schonfeld and Marie Peters for believing in me since almost day one in this world. You both are proof that childhood friendships are one of the most important things to nurture in life, after anthuriums. Ali, thank you for making me the ceramic pots featured throughout this book.

Thank you to all my friends and sister Allison for cheering me on and listening to me vent over group text since day one of this project. Sabrina Lockwood, thank you for lifting me up 24/7. I'm sorry we didn't get to feature all your half-dead houseplants in this book.

I'd like to also acknowledge Isabella Slowinski and Suzanne Kimpton for helping to take care of my plant and human babies while I wrote this book. It doesn't always take a village; sometimes it just takes two badass women.

Addis Café in Oak Park—home of the best lavender latte and avocado toast in the Midwest—thank you for cheering me on every week and providing me with a writing desk outside of my home.

Last but certainly not least, I'd like to acknowledge everyone who made this book a reality. Specifically, Jessica Walliser at The Quarto Group, you are a botanical book goddess. Thank you for believing in me and guiding me. Heather Godin and the art team at Quarto, thank you for your guidance and keen eye for aesthetics. Photographer Kat Schleicher, thank you for sharing your talent with me. You're a wizard behind the lens. Thank you to Esme Lintin, the illustrator, as well for her beautiful vivid drawings and putting up with the specificity of my requests!

This entire book is rooted in gratitude.

INDEX

Page numbers in **bold** indicate illustrations

mesocarp, 124
micropropagation, 30, 31, 129, 172
monstera
  adventitious roots, **73**
  aroid air layering, 104, 105, 107
  aroid mix, 56
  aroid propagation, 17
  bisexual (perfect) flowers, 25, 35
  chimeric variegation, 31
  humidity, 66
  inflorescences, 27, 35
  Monstera adansonii, 17
  Monstera deliciosa, 17, 84, 105
  Monstera deliciosa 'Albo-variega-
    ta,' 31, 61
  Monstera deliciosa 'Aurea,' **16**, **42**
  Monstera 'Devil Monster,' 100
  Monstera obliqua, 66, 84, 85
  node propagation, 84, 85
  petiole propagation, 88
  propagation boxes for, 66, 84
  propagation substrate, 54
  root rot, **42**
  spadix inflorescence, 27, 35
  tissue cultures, 129
  transitioning from water, 61
  water propagation and, 80
moonstones (*Pachyphytum ovi-
    ferum*), 142–143, **142**, **143**,
    144, **145**
mother plants
  air layering, 104, 105
  asexual propagation and, 30, **30**,
    33
  corms and, 172, 173
  definition of, 29
  dry succulent leaf propagation,
    136
  leaf vein propagation and, 92
  maintenance of, 11, 88
  pollination and, 114
  pups/offsets, 100, 150, 152
  regrowth, 78, 81
  root systems, 78, 174
  seed production and, 114
  simple layering and, 110, 112
  tissue culture and, 129
  variegation potential and, 31

N
natural rooting agents, 74, 75
nectar, 38

'New Guinea Ghost' (*Hoya nichol-
    soniae*)
  dead tissue, 181
  leaf status, 181–182
  pathogens, 181–182
  pest inspection, 181
  preparation, 181–183, **182**, **183**
  repotting, 185
  rooting hormone, 181
  root inspection, 182–183, **183**
  root regrowth, 185, **185**
  rot removal, 184, **184**
  semi-hydroponic, 185
  serenading, 182
  substrates for, 185
  supplies, 181
  water, 185
node propagation
  cutting selection, 85
  motivation for, 84
  potting, 86
  propagation boxes for, 84–86, **85**,
    86
nodes
  air layering, 104, **106**, 107
  butterflying method, 164, 165, 166,
    167
  cosmetic propagation, 180, **180**
  diagram, **14**
  function of, 15
  node stems, **16**
  simple layering, 110, 111, 112
  water propagation, 81, **81**, 82
nutrients
  air layering, 105
  anthurium pollination, 127
  corms and, 19, 172
  dry succulent leaf propagation,
    136
  earthworm castings, 49, **49**
  epiphytic plants, 57
  Fluval Stratum, 53, 185
  grafting, 161
  inorganic substrates and, 51, 53
  leaf blade propagation, 149
  nopales, 155
  oxygen and, 75
  peat moss, 48, **48**
  perlite, 49, **49**
  rooting hormone, 73
  semi-hydroponic propagation
    and, 50, 53
  senescence and, 181

simple layering, 110
soil succulent leaf propagation, 141
starting mix, 56
stems and, 15, 78
"straight to soil" propagation, 46
zygotes and, 39

O
offset propagation
  light for, 102
  potting, 102
  preparation, 101
  pup separation, 102
  rooting hormone, 100
  roots, 101, 102
  substrates for, 102
  supplies for, 100
  timeline for, 100
  water for, 102
onions, 18, **19**
Opuntia microdasys (bunny ear
    cactus), **154**, 155, **156**, **157**,
    158, **159**, **160**
Opuntia gosseliniana, **66**
Opuntia macrocentra (prickly pear
    cactus), 155, 159, **159**
organic substrates, 46, 47
outcrossing, 32, 34, 117
outlet timers, 64
ovaries, 25, 33, 34, 37, 128
oversaturation, 41
ovules
  fertilization of, 33, 35, 37, 39
  production of, 25, 39
Oxalis triangularis, 19, 20, **20**, **53**
oxygen
  anaerobic conditions, 41, 46
  aquarium pumps, 47, 61, 75, **75**
  photosynthesis, 15, 62
  substrates and, 61, 88
  water propagation and, 47, 82

P
Pachyphytum oviferum (moon-
    stones), 142–143, **142**, **143**,
    144, **145**
paintbrushes, 38, 42, **43**, 44, **44**,
    120, 122, **122**
PAR meters, 64
passive-hydroponic growing
  anthuriums and, 127
  Fluval Stratum and, 53
  hoya root-rot rehabilitation, 185